VOLUME FIFTY ONE

ADVANCES IN APPLIED MECHANICS

Advances in Crystals and Elastic Metamaterials, Part 1

VOLUME FIFTY ONE

ADVANCES IN
APPLIED MECHANICS

Advances in Crystals and Elastic Metamaterials, Part 1

Edited by

STÉPHANE P.A. BORDAS
Department of Computational Engineering Sciences, University of Luxembourg, Luxembourg and School of Engineering, Cardiff University, United Kingdom

DANIEL S. BALINT
Department of Mechanical Engineering, Imperial College London, London, United Kingdom

Volume Editor

MAHMOUD I. HUSSEIN
Ann and H.J. Smead Department of Aerospace Engineering Sciences, University of Colorado Boulder, Boulder, CO, United States

ELSEVIER

ACADEMIC PRESS
An imprint of Elsevier

Academic Press is an imprint of Elsevier
50 Hampshire Street, 5th Floor, Cambridge, MA 02139, United States
525 B Street, Suite 1800, San Diego, CA 92101-4495, United States
The Boulevard, Langford Lane, Kidlington, Oxford OX5 1GB, United Kingdom
125 London Wall, London, EC2Y 5AS, United Kingdom

First edition 2018

Notices

Knowledge and best practice in this field are constantly changing. As new research and experience broaden our understanding, changes in research methods, professional practices, or medical treatment may become necessary.

Practitioners and researchers must always rely on their own experience and knowledge in evaluating and using any information, methods, compounds, or experiments described herein. In using such information or methods they should be mindful of their own safety and the safety of others, including parties for whom they have a professional responsibility.

To the fullest extent of the law, neither the Publisher nor the authors, contributors, or editors, assume any liability for any injury and/or damage to persons or property as a matter of products liability, negligence or otherwise, or from any use or operation of any methods, products, instructions, or ideas contained in the material herein.

ISBN: 978-0-12-815100-6
ISSN: 0065-2156

Publisher: Zoe Kruze
Acquisition Editor: Jason Mitchell
Editorial Project Manager: Peter Llewellyn
Production Project Manager: Abdulla Sait
Cover Designer: Mark Rogers

Typeset by SPi Global, India

CONTENTS

CONTRIBUTORS

Hasan Al Ba'ba'a
Department of Mechanical and Aerospace Engineering, State University of New York at Buffalo, Buffalo, NY, United States

Clémence L. Bacquet
Ann and H.J. Smead Department of Aerospace Engineering Sciences, University of Colorado Boulder, Boulder, CO, United States

Michael J. Frazier
Department of Mechanical and Aerospace Engineering, University of California San Diego, San Diego, CA, United States

Mahmoud I. Hussein
Ann and H.J. Smead Department of Aerospace Engineering Sciences, University of Colorado Boulder, Boulder, CO, United States

Javier LLorca
IMDEA Materials Institute, Getafe; Department of Materials Science, Polytechnic University of Madrid, Madrid, Spain

Ricardo A. Lebensohn
Los Alamos National Laboratory, Los Alamos, NM, United States

Mostafa Nouh
Department of Mechanical and Aerospace Engineering, State University of New York at Buffalo, Buffalo, NY, United States

Javier Segurado
IMDEA Materials Institute, Getafe; Department of Materials Science, Polytechnic University of Madrid, Madrid, Spain

PREFACE

It is our pleasure to introduce a series of two volumes of *Advances in Applied Mechanics* focusing on recent contributions in crystals and elastic metamaterials.

Polycrystalline materials and homogenization approaches to describe their properties at the coarse, macroscopic engineering scale are topics of traditional importance in mechanics, which have been given remarkable attention especially within *Advances in Applied Mechanics*. We refer in particular to the work of Willis (1981) reviewing variational methods to compute overall properties of composites and of Castañeda and Suquet (1997) on nonlinear composite homogenization. This topic was revisited in a recent issue of this book series (Ostoja-Starzewski et al., 2016) discussing scaling to representative volume elements in random media, and homogenization theories with internal length scales (Aifantis, 2016).

Acoustic and elastic metamaterials are a topic that emerged in the last two decades bringing about a range of new extreme engineering properties and functionalities. Following the pioneering paper of Liu et al. (2000), the field has grown to produce artificial materials with simultaneously negative effective density and negative effective Young's modulus at subwavelength frequencies, that is, frequencies where the mechanical waves are much larger than the unit-cell feature sizes. The field then has advanced to produce novel applications such as sound focusing (Guenneau, Movchan, Pétursson, & Ramakrishna, 2007) and elastic cloaking (Milton, Briane, & Willis, 2006), to name a few. A recent review article provides an in-depth technical and literature review of the discipline (Hussein, Leamy, & Ruzzene, 2014).

In line with our strategy of combining contributions on theoretical, experimental, and computational approaches, this volume reviews recent advances in homogenization of polycrystals and contemporary concepts in engineering the dissipation in elastic metamaterials, with two contributions.

The first chapter deals with Computational Homogenization of Polycrystals, ranging from the digitalization of microstructures to the simulation of engineering components. Various approaches to computational homogenization are presented and compared in terms of computational expense and accuracy. The chapter also discusses approaches to build and compress statistically representative microstructures, including recent methods based on Wang tiles (Novák, Kučerová, & Zeman, 2012). Such approaches enable the determination of the size of representative volume elements (Doškářa, Zeman, Jarušková, & Nováka, 2018) and fuel the concept of jig-saw-design of metamaterial microstructures (Nežerka et al., 2018).

Homogenization is indeed equally important in the field of metamaterials as it is a means to calculate the long-wave effective properties. The second chapter presents Metadamping: Dissipation Emergence in Elastic Metamaterials. The dissipation in this context is not as is the case in microscopic relaxation mechanisms, due to defects, but is instead realized by exploiting local resonances at the unit-cell level and at macroscopic scales. This offers renewed flexibility in the design of new materials that retain their stiffness but offer significant damping properties as well.

The subsequent volume will constitute a sequel to the developments presented here by presenting two papers: one on phononic and photonics crystals for sensing applications and one on topological insulation in phononic crystals.

We look forward to this forthcoming issue and would like to thank all the authors of this volume for contributing their work and allowing a timely production of the two volumes.

Stéphane P.A. Bordas
Daniel S. Balint
Mahmoud I. Hussein

REFERENCES

Aifantis, E. C. (2016). Chapter one—Internal length gradient (ILG) material mechanics across scales and disciplines. *Advances in Applied Mechanics*, *49*, 1–110.

Castañeda, P. P., & Suquet, P. (1997). Nonlinear composites. *Advances in Applied Mechanics*, *34*, 171–212, 212a, 212b, 212c, 212d, 213–302.

Doškářa, M., Zeman, J., Jarušková, D., & Nováka, J. (2018). Wang tiling aided statistical determination of the representative volume element size of random heterogeneous materials. *European Journal of Mechanics—A/Solids*, *70*, 280–295.

Guenneau, S., Movchan, A., Pétursson, G., & Ramakrishna, S. A. (2007). Acoustic metamaterials for sound focusing and confinement. *New Journal of Physics*, *9*, 399.

Hussein, M. I., Leamy, M. J., & Ruzzene, M. (2014). Dynamics of phononic materials and structures: Historical origins, recent progress, and future outlook. *Applied Mechanics Reviews-Transactions of the ASME*, *66*(4), 040802.

Liu, Z., Zhang, X., Mao, Y., Zhu, Y. Y., Yang, Z., Chan, C. T., & Sheng, P. (2000). Locally resonant sonic materials. *Science*, *289*(5485), 1734–1736.

Milton, G. W., Briane, M., & Willis, J. R. (2006). On cloaking for elasticity and physical equations with a transformation invariant form. *New Journal of Physics*, *8*, 248.

Nežerka, V., Somr, M., Janda, T., Vorel, J., Doškář, M., Antoš, J., Zeman, J., & Novák, J. (2018). A jigsaw puzzle metamaterial concept. *Composite Structures*, *202*, 1275–1279.

Novák, J., Kučerová, A., & Zeman, J. (2012). Compressing random microstructures via stochastic Wang tilings. *Physical Review E*, *86*, 040104(R).

Ostoja-Starzewski, M., Kale, S., Karimi, P., Malyarenko, A., Raghavan, B., Ranganathan, S. I., & Zhung, J. (2016). Chapter two—Scaling to RVE in random media. *Advances in Applied Mechanics*, *49*, 111–211.

Willis, J. R. (1981). Variational and related methods for the overall properties of composites. *Advances in Applied Mechanics*, *21*, 1–78.

CHAPTER ONE

Computational Homogenization of Polycrystals

Javier Segurado*,†, Ricardo A. Lebensohn‡, Javier LLorca*,†,1
*IMDEA Materials Institute, Getafe, Madrid, Spain
†Department of Materials Science, Polytechnic University of Madrid, Madrid, Spain
‡Los Alamos National Laboratory, Los Alamos, NM, United States
[1]Corresponding author: e-mail address: javier.llorca@imdea.org

Contents

Advances in Applied Mechanics, Volume 51
ISSN 0065-2156
https://doi.org/10.1016/bs.aams.2018.07.001

Abstract

This paper reviews the current state of the art in the simulation of the mechanical behavior of polycrystalline materials by means of computational homogenization. The key ingredients of this modeling strategy are presented in detail starting with the parameters needed to describe polycrystalline microstructures and the digital representation of such microstructures in a suitable format to perform computational homogenization. The different crystal plasticity frameworks that can describe the physical mechanisms of deformation in single crystals (dislocation slip and twinning) at the microscopic level are presented next. This is followed by the description of computational homogenization methods based on mean-field approximations by means of the viscoplastic self-consistent approach, or on the full-field simulation of the mechanical response of a representative polycrystalline volume element by means of the finite element method or the fast Fourier transform-based method. Multiscale frameworks based on the combination of mean-field homogenization and the finite element method are presented next to model the plastic deformation of polycrystalline specimens of arbitrary geometry under complex mechanical loading. Examples of application to predict the strength, fatigue life, damage, and texture evolution under different conditions are presented to illustrate the capabilities of the different models. Finally, current challenges and future research directions in this field are summarized.

1. INTRODUCTION

Physically based modeling of the mechanical behavior of materials is an integral part of Materials Science and Engineering, as an indispensable tool to understand the relationship between microstructure and properties. However, the complexity of the deformation mechanisms that need to be represented, in addition to limitations in computational resources, have until recently restricted the goal of these formulations to provide qualitative trends or to complement phenomenological models that need to be adjusted from experimental data obtained under different conditions. Recent advances in simulation techniques, computational power, and multiscale modeling strategies are rapidly overcoming these limitations. These new paradigms are aimed at developing formulations that can quantitatively predict relationships between processing, microstructure and properties, and contribute to the design of materials with a given set of properties in silico, before they are manufactured in the laboratory. Successful examples of the latter can be found in the realm of metallic alloys (Allison, Li, & Wolverton, 2006; Olson, 2013) or structural composites (González, Vilatela, Molina-Aldareguía, Lopes, & LLorca, 2017; LLorca et al., 2011), and the roadmaps for implementation of these strategies in different industrial sectors are clearly delineated (Allison, Cowles, DeLoach, Pollock, & Spanos, 2013). Although the number of success stories is still limited, they show a dramatic reduction in the time necessary to develop new materials with optimized properties for specific applications, as well as a large return-of-investment. Thus, there is a manifest interest in both the scientific community and industry to improve and develop new multiscale modeling strategies.

Within this scenario, homogenization theory stands out as one of the most important tools to relate microstructure with effective properties in heterogeneous materials (Nemat-Nasser & Hori, 1993; Torquato, 2001). Homogenization theory assumes that the characteristic length scale of the microscopic domains (the average size of the heterogeneities) is much larger than the molecular dimensions (and, thus, continuum mechanics is applicable), and much smaller than the characteristic length scale of the macroscopic sample. Under these conditions, the macroscopic or effective properties of the material can be determined from geometrical features of the microstructure and properties of the different heterogeneities in the material. Following the pioneer work of Eshelby (1957), different homogenization mean-field formulations were developed first for linear materials (Hashin & Shtrikman, 1963; Hill, 1965b; Kröner, 1958; Mori & Tanaka, 1973). These methods were later extended to deal with nonlinear mechanical behaviors (nonlinear elasticity, plasticity, and viscoplasticity) by means of different linearization approximations (Berveiller & Zaoui, 1979; Hill, 1965a; Ponte Castañeda, 1991, 1996; Tandon & Weng, 1988).

Among these mean-field approaches, the elastic self-consistent (SC) scheme (Hershley, 1954) arises as the most reliable approach to obtain estimations of the elastic response of polycrystals (Hill, 1965b). This approximation assumes that each domain in a set that statistically represents the heterogeneous material is embedded in a homogeneous effective medium whose elastic properties are not known a priori, but need to be obtained as an average over those heterogeneities. The linear elastic SC formulation was extended to nonlinear viscoplastic polycrystals by Molinari, Canova, and Ahzi (1987a), taking into account that plastic deformation in each grain is accommodated by dislocation glide, resulting in local mechanical responses determined by the crystallography and the lattice orientation of each grain. This methodology provides a robust framework to establish physically based relationships between texture, microstructure, and mechanical properties of plastically deforming polycrystals, accounting for their evolving microstructure and the anisotropy of the single crystal grains. The viscoplastic self-consistent (VPSC) approximation was later improved by Lebensohn and Tomé (1993) and Lebensohn, Tomé, and Ponte Castañeda (2007), and the VPSC formulation is nowadays widely used to simulate the mechanical behavior of polycrystalline aggregates.

However, homogenization strategies based on mean-field approximations have two main limitations. The first one is that these approaches are based on a description of the microstructure based on average values of grain sizes, shapes, and orientations, and therefore cannot take into account the effects of local heterogeneities such as clustering of grains with different size, shape, or orientation, or to represent a given misorientation distribution. While these local microstructural features have often negligible effects on properties that depend on the average values of the micromechanical fields (like, for instance, the elastic modulus), this is not the case for properties that depend on the extremal values (e.g., damage) (LLorca & Segurado, 2004; Segurado, González, & LLorca, 2003; Segurado & LLorca, 2005). The second limitation arises from assuming that the micromechanical fields are constant in each phase/grain. Under these circumstances, it is difficult to simulate phenomena in which these fields are localized in narrow bands in one or several phases/grains, as a result of plastic deformation, damage progression, low strain rate sensitivity (i.e., high nonlinearity), strong mechanical contrasts, etc. (González & LLorca, 2007; González, Segurado, & LLorca, 2004; Totry, González, & LLorca, 2008). More sophisticated nonlinear homogenization methods have been developed in recent years, using linearization schemes at phase/grain level that also incorporate information on the second moments of the field fluctuations in each phase/grain, to ameliorate this second limitation, at the cost of more complex and expensive numerical algorithms (de Botton & Ponte Castañeda, 1995; Liu & Ponte Castañeda, 2004).

These limitations of mean-field homogenization approaches can be overcome by means of full-field homogenization. Using this strategy, the effective properties of the polycrystal are determined by means of the full-field solution of a boundary value problem of a microstructural representative volume element (RVE) under homogeneous boundary conditions. It has been established that the success and accuracy of computational homogenization approaches for polycrystals rely on the adequacy of three factors: the representation of the microstructure, the constitutive description of the single crystal behavior, and the numerical approach to solve the boundary value problem. From the RVE perspective, the first requirement is that the size of the RVE has to be large enough to provide an accurate statistical representation of the polycrystal's microstructure, as well as to lead to effective properties that are independent of the RVE dimensions. Obviously, the larger the RVE and the more complex the single crystal constitutive behavior, the higher the computational cost and, thus, fast and efficient numerical procedures to solve the boundary value problem are essential.

Full-field homogenization has been traditionally carried out using the finite element method (FEM) to solve the governing partial-differential equations (PDEs) of micromechanics, in combination with the CP formalism developed by Peirce, Asaro, and Needleman (1982). This model takes into account the process of plastic deformation by crystallographic slip and provides an accurate and physically based representation of this phenomenon at the microscopic level within each grain (Becker, 1991; Bhattacharyya, El-Danaf, Kalidindi, & Doherty, 2001; Delaire, Raphanel, & Rey, 2000; Miehe, Schrodër, & Schotte, 1999; Raabe, Sachtleber, Zhao, Roters, & Zaefferer, 2001). Alternatively, Moulinec and Suquet (1994) proposed a spectral formulation based on the efficient fast Fourier transform (FFT) algorithm to perform full-field homogenization of a periodic RVE. The governing PDEs were solved in this case by means of the computation of convolution integrals involving the periodic Green's function associated with the displacement (or velocity) field of a homogeneous linear reference material. This strategy was initially applied to composite materials (Eyre & Milton, 1999; Michel, Moulinec, & Suquet, 2000; Moulinec & Suquet, 1998), in which the source of heterogeneity is related to the spatial distribution of phases with different mechanical properties. It was later adapted to polycrystals, in which the heterogeneity is related to the spatial distribution of crystals with anisotropic mechanical properties (Lebensohn, 2001; Lebensohn, Brenner, Castelnau, & Rollett, 2008; Lebensohn, Castelnau, Brenner, & Gilormini, 2005; Lebensohn et al., 2009). It should also be noted that grain boundaries (GBs) of very different character are also ubiquitous in polycrystals and have to be modeled appropriately.

As a consequence of the above developments, simulation of the mechanical behavior of polycrystals has significantly improved in the last three decades and mature strategies based on either mean-field or full-field homogenization approaches are nowadays available. The state of the art of Computational Homogenization of Polycrystals (CHP) is summarized in this review, which is structured as follows. Section 2 is focused on the description of polycrystalline microstructures, including the experimental techniques available for their determination, and the digital representation of microstructures in a suitable format to perform the computational homogenization. Section 3 presents the crystal plasticity (CP) framework used by the different homogenization methods to describe the plastic deformation of each single crystal within the polycrystal. Both phenomenological and physically based models are described, as both are of interest for different applications. A detailed description of the Green's function-based VPSC homogenization approach is given in Section 4, while full-field homogenization strategies based on the numerical solution of a boundary value problem of an RVE using either the FEM or FFTs are presented in Sections 5.1 and 5.2, respectively. While a detailed description of CPFEM can be found elsewhere (Roters, Eisenlohr, Bieler, & Raabe, 2010; Roters, Eisenlohr, Hantcherli, et al., 2010) and, therefore, is omitted in this review, the efficient FFT-based formulation is described in detail. In particular, the underlying commonalities and differences between both Green's function-based CP formulations (VPSC and FFT-based models) are presented with a unified notation. Furthermore, Section 6 presents the formulation of multiscale frameworks for modeling the plastic deformation of polycrystalline specimens of arbitrary geometry under complex mechanical loading. These frameworks are based on the combination of polycrystal homogenization and FEM, by connecting the single crystal (microscopic) behavior with the specimen (macroscopic) response through an intermediate (mesoscopic) scale. The behavior at the mesoscopic scale is obtained from the homogenized response of a polycrystal that represents each (polycrystalline) material point in the FE mesh. A few applications of the described models are presented in Section 7, showing the capabilities of the different simulation strategies. These examples illustrate virtual design strategies, in which the mechanical response of polycrystalline microstructures can be accurately computed and optimized in silico before the material is actually manufactured. Finally, potential new applications of CHP are summarized in the last section, together with current limitations and topics that should be addressed in the future.

Throughout this chapter, vectors are denoted by bold lowercase roman letters $\mathbf{a}$, second-rank tensors by bold capital roman letters or bold Greek letters ($\mathbf{A}$ or $\boldsymbol{\sigma}$), and fourth-rank tensors by $\mathbb{A}$. In addition, bold capital roman ($\mathbf{E}$) and Greek letters ($\boldsymbol{\Omega}$) are used to express volume-average vectors, such as stress, strain, or velocities. A Cartesian coordinate system is used with respect to the orthonormal basis $(\mathbf{e}_1, \mathbf{e}_2, \mathbf{e}_3)$. The notations for tensor product, contraction, and double contraction products are $\mathbf{a}\otimes\mathbf{b} = a_i b_j \mathbf{e}_i \otimes \mathbf{e}_j$; $\mathbf{AB} = A_{ik}B_{kj}(\mathbf{e}_i \otimes \mathbf{e}_j)$; $\mathbf{a}\cdot\mathbf{b} = a_i b_i$ and $\mathbf{A} : \mathbf{B} = A_{ij}B_{ij}$. Finally $\mathbf{1}$ and $\mathbb{I}$ stand for the second- and fourth-rank identity tensors, respectively.

2. REPRESENTATION OF THE MICROSTRUCTURE

The first step to perform computational homogenization is the generation of synthetic RVEs of the microstructure. The definition of an RVE can be approached from two different perspectives. On the one hand, it can be defined as the minimum volume of the microstructure whose properties (obtained using either homogeneous traction or displacement boundary conditions) are equal to the effective properties of the heterogeneous solid. On the other hand, the RVE can be defined as the minimum volume of the microstructure whose statistical descriptors (that define the features of the heterogeneous microstructure) are equivalent to those found in the heterogeneous solid. Another important concept is that of statistical representative volume element (SRVE) (Jeulin, Kanit, & Forest, 2004). The SRVE does not fulfill one or both conditions imposed to the RVE by itself, but it does fulfill them in a statistical sense: the effective properties and statistical descriptors of the microstructure obtained by averaging the results obtained for many SRVE are equivalent to those obtained for an RVE. The use of SRVE is sometimes very useful in computational homogenization because the simulation of very large RVEs is computationally much more expensive than that of many smaller SRVEs.

2.1 Microstructural Parameters

Polycrystals are aggregates of single crystal grains, which are the building blocks of the RVE. Crystals are modeled as homogeneous polyhedra with well-defined crystallographic orientations separated by grain boundaries. In addition, internal discontinuities in the lattice of a grain due to the presence of twinned regions can also be included, due to the importance of these features to reproduce the mechanical response at the local level (for instance, in the nucleation of fatigue cracks). The main features to generate an RVE of a polycrystalline microstructure are the grain size and shape distribution, together with the orientation distribution function (ODF) that defines the crystallographic texture of the polycrystal. In addition, RVEs can include local microstructure heterogeneities or specific grain misorientation distributions.

Optical microscopy of polished sections of the microstructure is the standard technique to measure the grain size and shape distribution. Chemical etching can be used to reveal the grain boundaries and the statistical parameters that characterize the grain distribution can be automatically obtained using image-analysis software tools. Of course, the 2-D information obtained has to be transformed into 3-D and this operation may or may not be simple, depending on the microstructural features. For instance, there are algorithms available that can infer the 3-D grain size distribution from the data obtained in 2-D for equiaxed grain distributions (Heilbronner & Bruhn, 1998). Similarly, grain shapes can be accurately obtained from 2-D images in different orientations in the case of materials with well-defined anisotropy, as it is the case of rolled and extruded sheets. From the viewpoint of grain orientation, the ODF can be easily obtained from X-ray diffraction using texture goniometers. The information obtained can be directly used to model grain orientation distribution in the RVE.

More sophisticated methods have been developed in recent years to provide a more accurate quantification of the microstructure in polycrystals. Automated serial-sectioning by means of a microtome (Alkemper & Voorhees, 2001) or mechanical milling (Spowart, 2006; Spowart, Mullens, & Puchala, 2003) can be used in combination with software tools to build the 3-D microstructure and extract microstructural features, to determine grain size and shape distributions in 3-D. If higher resolution is required, sequential milling can be carried out using a focus ion beam or a femtosecond laser within a scanning electron microscope (Echlin, Husseini, Nees, & Pollock, 2011; Uchic, Groeber, Dimiduk, & Simmons, 2006). One additional advantage of this latter approach is that 3-D chemical and crystallographic information can be obtained during the process. In the particular case of polycrystals, the use of electron back-scattered diffraction (EBSD) provides information about the crystallographic orientation of each grain, the orientation relationship at the grain boundaries as well as the presence of twins within the grains (Fernández, Jérusalem, Gutiérrez-Urrutia, & Pérez-Prado, 2013; Konrad, Zaefferer, & Raabe, 2006; Uchic, Groeber, Wheeler, Scheltens, & Dimiduk, 2004). While serial-sectioning techniques are destructive, nondestructive characterization of the 3-D structure of polycrystals can be obtained by means X-ray tomography. Standard phase-contrast tomography provides information about porosity and second phases. Moreover, information about grain size and crystallographic orientation is presently available by means of diffraction-contrast tomography (Proudhon et al., 2015; Schmidt, Lauridsen, & Poulsen, 2008). Thus, detailed quantification of the microstructure of polycrystals is nowadays possible, but the acquisition time and the management and interpretation of the information contained in the massive datasets generated by 3-D characterization techniques may be problematic. Thus, the microstructural information necessary to build the RVE for computational homogenization should be assessed beforehand, to ensure efficient data collection and reconstruction.

It should be finally noted that the relevant microstructural information can be captured from 2-D or 3-D datasets by statistical descriptors, such as the 1-point and 2-point statistics (Torquato, 2001). 1-point statistics capture the probability associated with finding a specific phase or orientation at a point thrown randomly into the microstructure. They can be used to characterize the volume fraction of different phases or the texture of the polycrystal. 2-point statistics indicate the probability that both ends of a segment of a given length thrown randomly into the microstructure lie in the same phase. They can be used to characterize the spatial distribution, size, and shape of the different grains in the polycrystal. Once the particular statistical descriptors of a microstructure have been determined, phase-recovery algorithms can be used to generate RVEs, which are statistically equivalent, i.e., they share the same statistical descriptors (Doškář, Novák, & Zeman, 2014; Fullwood, Niezgoda, & Kalidindi, 2008; Niezgoda, Turner, Fullwood, & Kalidindi, 2010). The use of 2-point correlation function is particularly interesting for describing composites or two phase-materials. In the case of polycrystals (Niezgoda et al., 2010) each orientation should be considered as a different phase and the number of 2-point correlation functions to describe the microstructure becomes large. In this case, alternative microstructure descriptors such as the probability distribution of grain sizes or grain principal axis distribution are commonly used.

2.2 Digital Representation of the Microstructure

Once all the critical information about the microstructure of the polycrystal has been gathered, it is necessary to build a digital model of the RVE to perform computational homogenization. The RVEs can be constructed as a one-to-one representation of an actual microstructure measured from X-ray computed tomography or serial-sectioning data, or by generating synthetic microstructures from the statistical descriptors representing the microstructure. Two different types of discretizations are usually employed to define the geometry of the RVE. In the first one, the RVE is divided in voxels and grains are formed by groups of contiguous voxels that have the same crystallographic orientation (Fig. 1A). Voxel-based discretizations are the best option for direct representation of microstructures measured by three-dimensional techniques because they can be directly extracted from the measured data. In addition, voxel-based discretizations can be directly

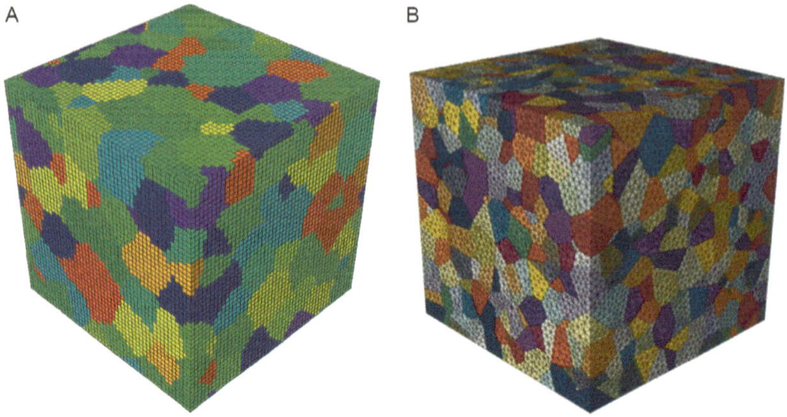

Fig. 1 Digital representations of an RVE of a polycrystal. (A) Voxel-based representation and (B) Voronoi-based representation.

exported into FFT-based codes as grid points or into FEM as regular hexahedral elements. Such voxel-based finite element meshes have good quality metrics and can be deformed up to very large strains (Segurado & LLorca, 2013). These voxel-based discretizations present, however, two major drawbacks. First, the grain boundaries are stepped surfaces and, thus, this type of representation is not appropriate to simulate phenomena localized at grain boundaries, like grain boundary sliding. Second, the voxel-based FEM discretization often leads to very large number of elements because the voxel size is controlled by the dimensions of the smallest features that have to be resolved within the RVE.

These limitations of the voxel-based digital models can be overcome with representations of the grain structure based on tessellation. A tessellation is a subdivision of the 3-D space into convex polyhedra that intersect only at their boundaries, which are flat surfaces. The most popular one is the Voronoi tessellation, which is generated by a set S of points in the three-dimensional space by assigning a volume $V_{\mathbf{x}_i}$ to the point $P_i(\mathbf{x}_i) \in S$ formed by all points $P(\mathbf{y})$ in the space which have P_i as their nearest neighbor (Redenbach, Shklyar, & Andrä, 2012). Cell boundaries in the Voronoi construction are, however, always equidistant from the generators of their cells and the range of cell patterns which can be generated is limited. Thus, Voronoi tessellations are not always suitable to reproduce the actual grain size distribution and weighted generalizations of the Voronoi model are frequently used. One possible generalization is the Laguerre tessellation in which the volume $V_{\mathbf{x}_i}$ in the space associated to the point $P_i \in S$ is formed by the points $P(\mathbf{y})$ that fulfill the condition

$$P(\mathbf{y}) \in V_{\mathbf{x}_i} \ \text{ if } \ d_L(\mathbf{y},\mathbf{x}_i) < d_L(\mathbf{y},\mathbf{x}_j), \ \ j \neq i \ \text{ and } \ \mathbf{x}_j \in S \tag{1}$$

where $d_L(\mathbf{y}, \mathbf{x}_i)$ is the "Laguerre" distance between points $\mathbf{y}$ and $\mathbf{x}_i$, which is given by

$$d_L(\mathbf{y},\mathbf{x}_i) = \| \mathbf{x}_i - \mathbf{y} \|^2 - r_i^2 \tag{2}$$

where r_i (>0) is the weight associated to point P_i. This definition leads to a partition of the space formed by convex, space-filling polyhedrons. It can be shown (Lautensack, 2007; Xue, Righetti, Telley, Liebling, & Mocellin, 1997) that if S is chosen as a system of nonoverlapping spheres characterized by the coordinates of their centers, $\mathbf{x}_i$ and the corresponding radius, r_i, each cell of Laguerre tessellation completely contains its generating sphere and the volume distribution of Laguerre cells is almost equal to volume distribution of their generating spheres.

Thus, the strategy to generate grain structures within the RVE begins with the experimental grain size distribution, which can be often approximated by a log-normal function. A polydisperse sphere distribution, following the experimental grain size distribution, is then introduced in the RVE and densely packed using collective rearrangement algorithms (Torquato, 2001) such as the force biased algorithm (Bargieł & Mościński, 1991). This algorithm starts with an initial distribution of spheres $S(\boldsymbol{x}_i, r_i)$ characterized by the position of the center $\boldsymbol{x}_i$ and the radius r_i distributed in the RVE. In this stage, overlapping of spheres is possible and allowed. Then, the algorithm attempts to reduce the overlaps between spheres by pushing apart overlapped spheres while small spheres are pushed to fill the empty spaces between large ones. After certain number of iterations, repositioning of overlapped spheres is stopped and the spheres gradually shrank to reduce the total amount of overlaps below a certain threshold. Finally, the coordinates of the centers of the spheres and their diameter are provided as output for the Laguerre tessellation. In other cases, Monte Carlo algorithms were used to obtain the spatial distribution of the set points for the tessellation, so the final cell size distribution coincides with the experimental grain size distribution (Cruzado et al., 2015; Mandal, Lao, Donegan, & Rollett, 2018).

Once the grain size distribution has been reproduced in the RVE, the crystallographic texture can be introduced by assigning different orientations to the grains to reproduce the statistical distribution given by the ODF. It should be noted that the minimum number of grains in the RVE should be large enough to reproduce accurately both the grain size distribution and the texture of the polycrystal. Moreover, more sophisticated grain orientation algorithms can be used to account for the presence of a given fraction of low- or high-angle grain boundaries, which may lead to important differences in the mechanical behavior at both microscopic and macroscopic levels.

The digital representation of the microstructure is an important and time-consuming task and microstructure builders have been developed, such as Neper (Neper, 2018) and Dream3D (Dream.3D, 2018). They also provide tools to clean up the voxelized microstructures obtained from tomography or serial-sectioning or from tessellation and to discretize the microstructure for full-field simulations.

3. CRYSTAL PLASTICITY MODELS

The first model that described the plastic deformation of metallic single crystals as a result of crystallographic slip was proposed by Taylor and Elam (1923, 1925). In this seminal work, the deformation of Al single crystals was analyzed and explained as the result of the shear deformation along 12 slip systems and the driving force for the shear deformation was the resolved shear stress on each slip system. A few years later, this model was used to analyze the deformation of a polycrystal as an aggregate of grains (Taylor, 1938). The ideas of Taylor for the deformation of single crystals were adapted into the framework of continuum mechanics by Hill (1966), in the case of small strains. The theory, based on the general internal variable thermodynamic formalism, was extended to finite deformations in the 1970s by Rice (1971) and Hill and Rice (1972). They used the concept of the multiplicative decomposition of the deformation gradient into elastic and plastic parts, introduced by Lee and Liu (1967). Different constitutive models based on this framework were developed for single crystals in the 1980s using either rate-independent formulations (Peirce, 1983; Peirce et al., 1982) or viscoplasticity (Asaro & Needleman, 1985). In parallel, attention was also paid to the development of rigorous numerical implementations of the models, including efficient and well-posed integration methods for the highly nonlinear viscoplastic laws (Cuitiño & Ortiz, 1992) and rigorous integration for the finite deformation framework (Miehe, 1996). The result of these studies—and many more not reviewed here—is a well-established theory of CP which will be summarized below.

The starting point of most CP models is the multiplicative decomposition of the deformation gradient $\mathbf{F}$ into its elastic ($\mathbf{F}^e$) and plastic ($\mathbf{F}^p$) parts (Fig. 2)

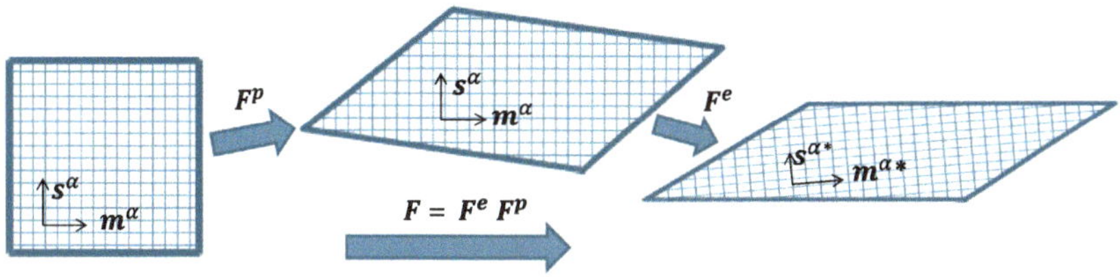

Fig. 2 Schematic of the multiplicative decomposition of the deformation gradient.

$$\mathbf{F} = \mathbf{F}^e \mathbf{F}^p \tag{3}$$

where the configuration defined by $\mathbf{F}^p$ is called the relaxed or intermediate configuration. In the context of CP, it is assumed that $\mathbf{F}^p$ leaves the crystal lattice undistorted and unrotated (Hill & Rice, 1972; Rice, 1971) and the rotation of the lattice is determined only by $\mathbf{F}^e$. Although this decomposition is generally accepted for CP, several issues as the existence and uniqueness of the decomposition and its connection with the microscopic distortion generated by the dislocations are still under debate (Reina & Conti, 2014; Reina, Schlmerkemper, & Conti, 2016).

From the definition of the velocity gradient, $\mathbf{L}$, expression (3) leads to

$$\mathbf{L} = \dot{\mathbf{F}}\mathbf{F}^{-1} = \dot{\mathbf{F}}^e \mathbf{F}^{e^{-1}} + \mathbf{F}^e \dot{\mathbf{F}}^p \mathbf{F}^{p^{-1}} \mathbf{F}^{e^{-1}} \tag{4}$$

where $\mathbf{L}^p = \dot{\mathbf{F}}^p \mathbf{F}^{p^{-1}}$ stands for the plastic deformation rate in the intermediate configuration.

The constitutive equations can be obtained from the energy density per unit volume expressed in the intermediate configuration (Cuitiño & Ortiz, 1992; Han, Gao, Huang, & Nix, 2005b). Following the internal variable formalism, the free energy density, ψ, can be written as

$$\psi = \psi(\mathbf{F}; \mathbf{F}^p, \mathbf{q}) \tag{5}$$

where $\mathbf{q}$ is a set of internal variables. The free energy density can be split into the elastic and plastic energy densities (Cuitiño & Ortiz, 1992),

$$\psi = \psi^e\left(\mathbf{F}\mathbf{F}^{p^{-1}}\right) + \psi^p(\mathbf{F}^p, \mathbf{q}). \tag{6}$$

and the second Piola–Kirchhoff stress tensor in the intermediate configuration $\mathbf{S}$ can then be obtained as

$$\mathbf{S} = \frac{\partial \psi^e}{\partial \mathbf{E}^e} \tag{7}$$

where $\mathbf{E}^e$ stands for the Green–Lagrange elastic strain in the intermediate configuration and is given by

$$\mathbf{E}^e = \frac{1}{2}(\mathbf{F}^{e^T}\mathbf{F}^{\mathbf{e}} - \mathbf{I}). \tag{8}$$

The second Piola–Kirchhoff stress in the intermediate configuration $\mathbf{S}$ is related with the Cauchy stress, $\boldsymbol{\sigma}$, according to

$$\mathbf{S} = \frac{1}{J}\mathbf{F}^{e^{-1}}\boldsymbol{\sigma}\mathbf{F}^{e^{-T}} \tag{9}$$

where J is the determinant of $\mathbf{F}$. Assuming a quadratic potential for the elastic energy ψ^e in Eq. (6), $\mathbf{S}$ can be expressed as a linear function of the Green–Lagrange elastic strain according to

$$\mathbf{S} = \mathbb{L}\mathbf{E}^e \tag{10}$$

where $\mathbb{L}$ stands for the fourth-order elastic stiffness tensor of the single crystal.

The crystallographic nature of the plastic deformation of single crystals is described by two orthogonal unit vectors which define the slip system k, $\mathbf{s}^k$, and $\mathbf{m}^k$ that stand for the slip direction and slip plane normal, respectively, and that remain invariant in the intermediate configuration (Fig. 2). The available slip systems of a given crystalline material $k = 1, 2, \ldots, n$ are determined by its lattice. For instance, FCC materials presented 12 slip systems characterized by the $\{111\}$ planes and the $\langle 110\rangle$ directions.

Rice (1971) proposed the conventional flow rule of a single crystal based on this geometrical description according to

$$\mathbf{L}^p = \sum_k \dot{\gamma}^k \mathbf{s}^k \otimes \mathbf{m}^k = \sum_k \dot{\gamma}^k \mathbf{Z}^k \tag{11}$$

where γ^k are internal variables of the model ($\gamma^k \in \mathbf{q}$) that account for the accumulated plastic slip in each slip system k. The dyadic product $\mathbf{s}^k \otimes \mathbf{m}^k$ is the nonsymmetric Schmid tensor of the system k, $\mathbf{Z}^k$.

The evolution of γ^k is normally assumed to be dependent on the internal variables chosen to describe the state $\mathbf{q}$ and on the stress through the resolved shear stress τ^k

$$\dot{\gamma}^k = \dot{\gamma}^k(\tau^k;\mathbf{q}). \tag{12}$$

The resolved shear stress τ^k is defined in the framework of finite strains as (Cuitiño & Ortiz, 1992)

$$\tau^k = \mathbf{F}^{e\,T}\mathbf{F}^e\mathbf{S} : \mathbf{s}^k \otimes \mathbf{m}^k \tag{13}$$

which can be simplified in the usual case of small elastic strains ($\mathbf{F}^{e^T}\mathbf{F}^e \approx \mathbf{I}$) as

$$\tau^k \approx \mathbf{S} : \mathbf{s}^k \otimes \mathbf{m}^k = \mathbf{S} : \mathbf{Z}^k \tag{14}$$

The last two ingredients of a CP model are the particular functions that dictate the shear rate $\dot{\gamma}^k$ of each slip system (Eq. 12) as well as the evolution of the internal variables $\mathbf{q}$ during deformation. Many different flavors of CP have been developed in the last decades depending on these functions and the most relevant ones will be reviewed below grouped in three categories: phenomenological, physically based and strain gradient plasticity models.

3.1 Phenomenological Crystal Plasticity Models

Crystal plasticity, in contrast with classical macroscopic plasticity, has a clear physical basis and always includes explicit microscopic information of the material, such as the geometrical definition of the active slip systems. However, CP models can use phenomenological expressions to define both the slip rates and the evolution of the internal variables. This phenomenological approach is based on the classical constitutive equation theory, and the internal variables that determine the state of the crystal, as well as their evolution laws, are not directly related to the microscopic physical magnitudes or processes. The first approaches (Hill & Rice, 1972; Rice, 1971) were rate-independent but they were relatively complex and prone to convergence problems because multiple combinations of shear increments can lead to the same plastic strain (Peirce et al., 1982). Although most of the problems associated with rate-independent formulations have been overcome (Anand & Kothari, 1996), viscoplastic formulations became a very popular alternative since they were introduced by Peirce, Asaro, and Needleman (1983).

The set of internal variables, $\mathbf{q}$, contains the information about the accumulated plastic slip, γ^k, and the critical resolved shear stress (CRSS), g^k, in each slip system. For a given state, g^k corresponds to the minimum value of the resolved shear stress, Eq. (13), to activate the plastic flow in that system. The slip rate, Eq. (12), is given by a power-law function according to

$$\dot{\gamma}^k = \dot{\gamma}_0 \left(\frac{|\tau^k|}{g^k}\right)^{1/m} \mathrm{sign}(\tau^k) \tag{15}$$

where $\dot{\gamma}_0$ and m stand for the reference strain rate and the strain rate-sensitivity parameter, respectively. The strain rate-sensitivity parameter is equivalent to $1/n$, where n stands for the strain rate-sensitivity exponent. Note that only positive values of $\dot{\gamma}^k$ were considered in the original work of Peirce et al. (1983) by including positive and negative slip directions $\pm\mathbf{s}^k$ while in Eq. (15), as in most of the present formulations, $\dot{\gamma}^k$ can take both signs and only one direction of $\mathbf{s}^k$ is considered.

Regarding strain hardening, the model by Peirce et al. (1983) was developed for monotonic loading and therefore only considered isotropic hardening. The CRSS on each system included the explicit contributions of the slip in the same system (self-hardening) and of the slip on all the other slip systems (latent hardening) and the evolution of the CRSS can be written as

$$\dot{g}^k = \sum_j h_{kj}|\dot{\gamma}^j| \tag{16}$$

where h_{kj} are the latent hardening moduli, with h_{kk} the self-hardening modulus. The model proposed by Peirce et al. (1983) defined the hardening modulus, Eq. (16), as

$$h_{kj} = h(\Gamma)[q + (1-q)\delta_{kj}] \tag{17}$$

where q is a parameter defining the latent hardening, Γ is the accumulated shear on all the slip systems

$$\Gamma = \sum_k \int |\dot{\gamma}^k| dt \tag{18}$$

and the function $h(\Gamma)$ proposed for FCC single crystals takes the form

$$h(\Gamma) = h_0 \operatorname{sech}^2\left(\frac{h_0\Gamma}{\tau_s - \tau_0}\right) \tag{19}$$

where h_0 is the initial hardening modulus and τ_0 and τ_s stand for the initial and the saturation values of CRSS, respectively.

Alternative expressions of $h(\Gamma)$ have been proposed to account for nonvanishing hardening rates at large plastic strains. This is the case of the Voce hardening law proposed in Tomé, Canova, Kocks, Christodoulou, and Jonas (1984) that defines h as

$$h(\Gamma) = h_s\left[1 - \exp\left(\frac{-h_0\Gamma}{\tau_s - \tau_0}\right)\right] + (\tau_s - \tau_0 + h_s\Gamma)\frac{h_0}{\tau_s - \tau_0}\exp\left(\frac{-h_0\Gamma}{\tau_s - \tau_0}\right) \tag{20}$$

where a new parameter h_s is introduced to define the hardening slope at large plastic strains. An alternative hardening model, developed by Bassani and Wu (1991), proposed a similar nonvanishing hardening rate through a parameter h_s using Eq. (19) as starting point. The corresponding hardening moduli can be expressed as

$$h_{kk} = \left[(h_0 - h_s)\mathrm{sech}^2\left(\frac{h_0 - h_s}{\tau_s - \tau_0}\gamma_k\right) + h_s\right]\left[1 + \sum_{j\neq k} f_{kj}\tanh\left(\frac{\gamma_j}{\gamma_k}\right)\right] \tag{21}$$
$$h_{kj} = q h_{kk}$$

where h_{kk} and h_{kj} stand for the self-hardening and the latent hardening coefficients in Eq. (16). In contrast to Peirce et al. (1983), the dependence of the self-hardening coefficient on the slip accumulated in the different slip systems is not uniform in this formulation, but given by some interaction coefficients f_{kj} that depend on the nature of the dislocation junctions between the slip systems. Most of the phenomenological CP models developed since then for monotonic loading are based on either of the three hardening models summarized above.

The extension of this phenomenological framework to account for cyclic deformation is done by the introduction in the constitutive model of the effect of a backstress to formulate kinematic hardening laws at the crystal level (Cailletaud, 1992; Hu, Rauch, & Teodosiu, 1992; Méric, Poubanne, & Cailletaud, 1991) and similar approaches have been used to model creep (Hasija, Ghosh, Mills, & Joseph, 2003; Venkatramani, Ghosh, & Mills, 2007). The plastic shear rate depends in this case on both the CRSS and the backstress according to (Cailletaud, 1992)

$$\dot{\gamma}^k = \begin{cases} \dot{\gamma}_0\left(\dfrac{|\tau^k - \chi^k| - g^k}{K}\right)^{1/m} \mathrm{sign}(\tau^k - \chi^k) & \text{if } |\tau^k - \chi^k| \geq g^k \\ 0 & \text{if } |\tau^k - \chi^k| < g^k \end{cases} \tag{22}$$

where χ^k stands for the backstress of system k and K is a numerical parameter.

The kinematic hardening is defined by the evolution of the backstress and Cailletaud (1992) proposed the following expression for FCC metals

$$\dot{\chi}^k = c\dot{\gamma}^k - d|\dot{\gamma}^k| \tag{23}$$

where c and d are material parameters that define the hardening rate.

Other expressions for nonlinear phenomenological kinematic hardening laws at the crystal level have been proposed based in most cases in well-established relations at the macroscale. One of the most common expressions is the Frederick–Armstrong law, adapted from macroscopic plasticity to the crystal level in Méric et al. (1991) according to

$$\dot{\chi}^k = c\dot{\gamma}^k - d|\dot{\gamma}^k|\chi^k. \tag{24}$$

to account for the Bauschinger effect in single crystal Ni-based superalloys. More recently, Cruzado, LLorca, and Segurado (2017) presented a phenomenological CP model for cyclic loading that includes the effect of cyclic softening and an alternative evolution of the backstress to account for the mean stress relaxation. The plastic slip rate in this model is expressed as

$$\dot{\gamma}^k = \dot{\gamma}_0 \left(\frac{|\tau^k - \chi^k|}{g^k} \right)^{1/m} \operatorname{sign}(\tau^k - \chi^k). \tag{25}$$

The evolution of the backstress is obtained as a simplified version of the Ohno–Wang macroscopic model (Ohno & Wang, 1993) limited to the first two terms

$$\dot{\chi}^k = c\dot{\gamma}^k - d\chi^k|\dot{\gamma}^k| \left(\frac{|\chi^k|}{c/d} \right)^r \tag{26}$$

where c and d are the parameters of the Frederick–Armstrong model while r is an extra parameter that controls the mean stress relaxation velocity. Finally, the cyclic softening was accounted for through a new internal variable, the cyclic plastic slip, γ_{cyc}, is given by,

$$\gamma_{cyc} = \sum_k \int_0^t |\dot{\gamma}^k| dt - \sum_k \left| \int_0^t \dot{\gamma}^k dt \right| \tag{27}$$

which was taken into account in the evolution of the CRSS, $\dot{g}^k = \dot{g}^k(\gamma_{cyc})$, using a Voce type law with negative slope.

3.2 Physically Based Crystal Plasticity Models

Physically based CP models (as opposed to phenomenological ones) contain a stronger physical connection with the microscopic mechanisms of plastic deformation. Thus, some microscopic physical quantities are included as internal variables (i.e., dislocation densities) and/or rate equations are based on the active microscopic deformation mechanisms. In addition, these models usually include as input some additional microstructure information such as the initial dislocation density, the volume fraction of precipitates or second phases and, very often, include the effect of temperature.

The first ingredient of physically based models is the relationship between the plastic slip rate and the dislocation movement. This is introduced through the Orowan equation (Orowan, 1934) that connects the plastic slip rate on a given slip system, $\dot{\gamma}^k$, with the mobile dislocation density, ρ_m^k, the Burgers vector, b^k, and the average dislocation velocity, $\overline{v}^k$, according to

$$\dot{\gamma}^k = \rho_m^k b^k \overline{v}^k. \tag{28}$$

This equation replaces the power-law rate equations in the phenomenological rate-dependent models, i.e., Eqs. (15), (22), (25). The driving force for the dislocation movement in Eq. (28) is the resolved shear stress, Eq. (13), that is introduced through the average velocity. In metals with compact lattices (FCC and for some slip systems in HCP), lattice friction is negligible and slip occurs at very low stresses. In this case, a linear viscous relation can be postulated between the resolved stress and the velocity of a single dislocation (Hirth & Lothe, 1982) leading to

$$v^k = \frac{\tau^k b^k}{B} \tag{29}$$

where B is the drag coefficient, a material parameter that depends on temperature. If the density of the dislocations is very low, the average dislocation velocity in Eq. (28) can be obtained directly from the CRSS using Eq. (29), and a linear viscous relationship is established between the CRSS and the plastic slip rate. However, this relationship is not realistic in most cases because dislocations have to overcome different barriers during slip, leading to an average dislocation velocity different from the one given by Eq. (29).

The barriers to dislocation movement can be classified as temperature-dependent (thermal) or independent (athermal), depending on whether thermal activation can help to overcome the obstacle. For instance, long-range elastic interactions among dislocations introduce an athermal threshold CRSS for the dislocation movement τ_a. Taylor (1938) determined this threshold stress for pure metals, which is given by

$$\tau_a \propto \mu b \sqrt{\rho} \tag{30}$$

where μ is the shear modulus and $1/\sqrt{\rho}$ stands for the average distance between dislocations.

The thermal barriers are due to the short-range interactions of dislocations with other dislocations (jogs created by the intersection of forest dislocations and their movement by vacancy generation) and point defects. The strength of the barrier at 0 K is given by τ_t and the CRSS necessary to overcome the barrier is given by $\tau_a + \tau_t$. At finite temperatures, thermal energy helps the dislocation to jump over the barrier, and the average dislocation velocity (Eq. 28) becomes dependent on the temperature.

The influence of temperature on dislocation slip under short-range interactions was studied in detail in Kocks (1975) with the framework of the transition state theory and this work is the basis of most temperature-dependent physically based CP models in the literature (Cheong & Busso, 2004; Kothari & Anand, 1998; Ma & Roters, 2004; Rodríguez-Galán, Sabirov, & Segurado, 2015; Shahba & Ghosh, 2016). In summary, the average dislocation velocity to be inserted in Eq. (28) can be written as

$$\bar{v}^k = \begin{cases} 0 & \text{if} \quad 0 \leq |\tau^k| \leq \tau_a \\ \bar{l}^k \nu_0 \exp\left(-\dfrac{\Delta G}{kT}\right) & \text{if} \quad \tau_a < |\tau^k| < \tau_a + \tau_t \end{cases} \tag{31}$$

where $\bar{l}^k$ is the average distance between the obstacles in the slip system k, ν_0 the attempt frequency, k the Boltzmann constant, and T the absolute temperature. $\Delta G(\tau^k)$ stands for the Gibbs free energy that has to be supplied by thermal fluctuations to overcome the obstacle, which depends on the applied shear stress τ^k, and the exponential term expresses the probability of the occurrence of a jump over a short-range barrier.

The evolution of the Gibbs free energy with the applied shear stress in the presence of a general array of obstacles in the slip plane can be expressed as (Kocks, 1975),

$$\Delta G(\tau^k) = \Delta F\left[1 - \left\langle \frac{|\tau^k| - \tau_a}{\tau_t} \right\rangle^p\right]^q \tag{32}$$

where ΔF is the activation free energy necessary to overcome the obstacles without the aid of an applied shear stress, and $\langle x \rangle$ stand for the Macaulay brackets, which return x if $x > 0$ and 0 otherwise. Finally, p and q (in the range $0 \leq p \leq 1$ and $1 \leq q \leq 2$) are two parameters that define the strength of the obstacle as a function of the distance propagated by the dislocation. A simplification of Eq. (32) often found in CP models (Dunne, Rugg, & Walker, 2007; Ma & Roters, 2004) assumes that the obstacle strength is constant ($p = q = 1$) and Eq. (32) can be written as

$$\Delta G(\tau^k) = \Delta F - \tau^\alpha V \tag{33}$$

where V is the activation volume, a material constant that determines the actual volume of the material affected by the short-range dislocation-obstacle interaction.

It should be finally noted that the most important contribution to the thermal strength, τ_t, in metals with noncompact lattices (such as BCC or low density planes in HCP crystals) is the lattice friction. In this case, τ_t at 0 K is the Peierls stress that can be obtained from the simple model by Peierls (1940) and Nabarro (1947) or from atomistic or ab initio simulations (see, for instance, Stukowski, Cereceda, Swinburne, & Marian, 2015 for BCC W and Yasi et al., 2009 for pyramidal slip in HCP Mg).

In addition to a physically based model of $\dot{\gamma}^k$, many physically based CP models incorporate micromechanical internal variables, such as dislocation densities, to account directly for the strain hardening from physical considerations. The relationship between dislocation densities and the athermal strength is based on the Taylor model, Eq. (30). This expression is usually enriched by including several terms that account for the interactions between dislocations to obtain the flow stress in the different slip systems (here denoted as g^k to emphasize the relation with the CRSS introduced in the phenomenological models). In the case of FCC crystals, this can be expressed as (Franciosi & Zaoui, 1982)

$$g^k = \mu^k b \sqrt{\sum_j a_{kj} \rho^j} \tag{34}$$

where ρ^j is dislocation density in the slip system j and a_{kj} are a set of nondimensional coefficients that determine the self-hardening and the latent hardening due to the interactions among dislocations (similar to the phenomenological expression, Eq. (16)). The values a_{kj} for different crystal lattices can be obtained from dislocation dynamics simulations (Bertin, Tomé, Beyerlein, Barnett, & Capolungo, 2014; Devincre, Hoc, & Kubin, 2008).

The dislocation densities in Eq. (34), ρ^k, are introduced as local internal variables and correspond to the average of dislocation length per unit volume at each point of the crystal. The evolution of these internal variables is normally based in the Kocks–Mecking model (Estrin & Mecking, 1984; Mecking & Kocks, 1981). This model considers that hardening is controlled by the competition between storage and annihilation of dislocations and that both processes are additive. Thus, the corresponding evolution law for one slip system k is given by

$$\dot{\rho}^k = \left[\frac{1}{\ell(\rho^k)} - 2 y_c \rho^k \right] |\dot{\gamma}^k|. \tag{35}$$

where the first term, $1/\ell(\rho^k)$, is athermal and controls the storage of dislocations and $\ell(\rho^k)$ is the dislocation mean free path, which corresponds to the distance traveled by a dislocation segment before it is stopped by an obstacle. In the absence of precipitates or other obstacles to the dislocation motion, $\ell(\rho^k) = k_1/\sqrt{\rho^k}$, where k_1 is a material constant. The second term, $2y_c\rho^k$, is associated with the dislocation annihilation due to dynamic recovery. It depends on the temperature and is characterized by y_c, a constant that depends on the critical annihilation distance between dislocations. It should be noted that Eq. (35) can also be extended to alloys with a distribution of impenetrable obstacles. In this case, the dislocation mean free path will be determined by the obstacle spacing (Estrin & Mecking, 1984).

The generalization of the Kocks–Mecking model for multiple slip systems reads

$$\dot{\rho}^k = \left[\frac{1}{\ell^k(\rho^1,\ldots,\rho^n)} - 2y_c\rho^k\right]|\dot{\gamma}^k| \tag{36}$$

and

$$\ell^k \approx \frac{k_1}{\sqrt{\sum_{j\neq k}\rho^k}}. \tag{37}$$

Several modifications of the Kocks–Mecking model can be found in the literature based on the results obtained from simplified dislocation mechanics models (Cheong & Busso, 2004) or dislocation dynamics simulations (de Sansal, Devincre, & Kubin, 2010; Devincre et al., 2008). Moreover, the introduction of the distance to grain boundary as an upper bound to the dislocation mean free path has been successfully used to simulate the effect of grain size on the strength of Cu polycrystals (Hauoala, Segurado, & LLorca, 2018).

3.3 Strain Gradient Crystal Plasticity Models

The CP framework presented above is local, i.e., the material response at a given point depends only on the local values of both state and internal variables at that point. Therefore, the constitutive equations are size-independent because there are not length scales involved. However, the experimental evidence and dislocation dynamics simulations show that the strength of single crystals is size-dependent when they are subjected to homogeneous (El-Awady, 2015) and inhomogeneous plastic deformation, such as nanoindentation (Sánchez-Martín et al., 2014a; Stelmashenko, Walls, Brown, & Milman, 1993) or bending tests of single crystal cantilever beams (Gong & Wilkinson, 2011; Kiener, Grosinger, Dehm, & Pippan, 2008; Motz, Schoeber, & Pippan, 2005). The development of plastic strain heterogeneities in the crystal deformation is also the reason behind the well-known dependency of the plastic strength of polycrystals on grain size (Hall, 1951; Petch, 1953).

The effect of the plastic heterogeneities in the mechanical response of a crystal was first rationalized by Nye (1953) and Ashby (1970) as a result of the interaction between statistically stored dislocations (SSDs), which evolve from random trapping processes during plastic deformation, and geometrically necessary dislocations (GNDs) induced by the presence of plastic strain gradients. From the modeling viewpoint, the most common approach to account for the effect of plastic strain heterogeneities in the material response is by introducing the influence of some plastic strain gradient measure in the constitutive equation, leading to the so-called strain gradient plasticity. This idea was introduced by Aifantis (1987) for macroscopic plasticity and further developed to account for size effects in polycrystals (Fleck, Muller, Ashby, & Hutchinson, 1994; Nix & Gao, 1998). This modeling framework was extended to CP by Acharya and Bassani (1995) and Shu and Fleck (1999), who in these seminal papers introduced the two different strategies followed in the strain gradient crystal plasticity (SGCP) models developed since then.

In the first modeling approach, *lower-order* SGCP, strain gradients enter only in the instantaneous hardening moduli, while the thermodynamic consistency is preserved (Acharya, Bassani, & Beaudoin, 2003). This approximation allows the use of the classical mathematical framework of boundary value problems in standard plasticity (Acharya & Bassani, 1995; Busso, Meissonnier, & O'Dowd, 2000; Cheong, Busso, & Arsenlis, 2005; Dai & Parks, 1997; Dunne et al., 2007; Han et al., 2005b; Han, Gao, Huang, & Nix, 2005a; Ma, Roters, & Raabe, 2006) The second approach stands for the *higher order* SGCP models in which some internal variables are chosen as kinematic variables. This implies the introduction of stresses conjugated to these kinematic variables as well as the corresponding "higher order" boundary conditions (Bardella, 2006; Bardella, Segurado, Panteghini, & Llorca, 2013; Bayley, Brekelmans, & Geers, 2006; Borg, Niordson, & Kysar, 2008; Evers, Brekelmans, & Geers, 2004; Gurtin, 2002, 2008; Gurtin & Needleman, 2005; Niordson & Kysar, 2014; Yefimov, van der Giessen, & Groma, 2004).

The obvious benefit of lower-order formulations is their simple structure, so they can be easily implemented in existing general-purpose finite element codes. In addition, they avoid the additional higher order boundary conditions, which are not always easy to interpret physically. Due to these facts, lower-order SGCP models are the basis of most physically based CP models that account for size effects. In addition, they have been coupled to vacancy diffusion models to account for the effect of dislocation climb, a critical process to simulate creep deformation (Geers et al., 2014). On the other hand, the main limitation of lower-order formulations is the impossibility of accounting for boundary layers in constrained plastic flow, as the development of these layers requires the use of higher order boundary conditions on the plastic slip fields. In this sense, higher order models are capable of accounting for the gradient development due to the presence of a passivation layer or a grain boundary within a polycrystal.

3.3.1 Lower Order SGCP

The starting point in lower-order SGCP models is the definition of a measure for the plastic strain gradients. This is usually based on the relationship between the plastic slip gradients and the density of GNDs. For single slip, the plastic slip gradient gives rise to the development of a density of GNDs, ρ_{GND}, to maintain continuity in the crystal according to (Nye, 1953)

$$\rho_{\text{GND}} = -\frac{1}{b}\frac{\partial \gamma}{\partial x_n} \tag{38}$$

where γ is the plastic slip, b the Burgers vector and x_n stands for the coordinate normal to the slip direction. The generalization of the previous equation to multiple slip systems is expressed through the Nye's tensor, $\boldsymbol{\alpha}$, which was introduced by Nye (1953) and generalized by Arsenlis and Parks (1999) as

$$\alpha_{ij} = \sum_a \rho^a_{\text{GND}} b^a_i t^a_j \tag{39}$$

where a stands for a straight dislocation segment of length l^a parallel to $\mathbf{t}^a$ with Burgers vector $\mathbf{b}^a$, and $\rho^a{}_{\text{GND}}$ stands for the length of segment a per unit volume. Nye's tensor accounts for the lattice curvature and, therefore, can be expressed as function of the plastic slip gradients in the case of small strains according to (Arsenlis & Parks, 1999)

$$\boldsymbol{\alpha} = \nabla \times \sum_\alpha \gamma^k \mathbf{s}^k \otimes \mathbf{m}^k \tag{40}$$

where $\nabla\times$ represents the curl operator. In the case of finite strains, the Nye tensor is usually defined as (Busso et al., 2000; Ma et al., 2006)

$$\boldsymbol{\alpha} = \nabla \times \mathbf{F}^{\mathrm{p}} \tag{41}$$

although other tensorial measures of the plastic incompatibility are defined in the literature, as the finite strain geometric dislocation tensor **G** proposed in Cermelli and Gurtin (2001) and given by

$$\mathbf{G} = \mathbf{F}^{\mathrm{p}}(\nabla \times \mathbf{F}^{\mathrm{p}}) \tag{42}$$

In lower-order gradient models, the plastic strain gradient is included in the hardening expression. This is normally done adding to the Taylor hardening model (Eq. 34) that relates the CRSS with the dislocation density, a term that accounts for the GND density and depends on some plastic gradient measure. Along these lines, a first model inspired on physical considerations, was presented by Han et al. (2005b, 2005a), who defined a GND density for each slip system, η^k, based in a projection of the dislocation tensor (Eq. 42) in the system. Then, the effective gradient-dependent CRSS, g_{eff}, was given by

$$g_{eff}^k = g_0\sqrt{(g^k/g_0) + l\eta^k} \tag{43}$$

where g^k is the CRSS of the system k in the absence of gradients and l stands for an intrinsic length scale. It should be noted that Eq. (43) resembles the phenomenological approaches of macroscopic gradient plasticity developed by Nix and Gao (1998) but the physical origin of l is clearer and can be obtained from physical considerations.

If the density of GNDs in each system is explicitly resolved, the CRSS can be defined as (Cheong et al., 2005)

$$g^k = \mu b\sqrt{\sum_j \left(a_{kj}^{\mathrm{SSD}}\rho_{\mathrm{SSD}}^j + a_{kj}^{\mathrm{GND}}\rho_{\mathrm{GND}}^j\right)} \tag{44}$$

where ρ_{SSD}^j stands for the SSDs density and ρ_{GND}^j corresponds to the GNDs density in system j. The coefficients a_{kj}^{SSD} and a_{kj}^{GND} define the latent hardening interactions among the different slip systems. The distribution of GND densities on the different systems ρ_{SSD}^k is not uniquely determined by the dislocation tensor and additional constraints have to be imposed. Arsenlis and Parks (1999) proposed to minimize either the dislocation density or the dislocation length to obtain the actual GND distribution. Alternatively, the GND evolution can be determined by integrating some evolution laws obtained by expressing the Nye tensor in terms of the spatial gradient of the slip rate (Busso et al., 2000; Cheong et al., 2005; Dunne et al., 2007; Ma et al., 2006).

In order to implement these models in a finite element framework, the slip gradients or Nye tensor at the integration point level have to be determined. This task has been traditionally done using a very efficient local element approach (Busso et al., 2000; Dunne et al., 2007) in which the internal variables included in the gradient term are extrapolated from the integration points to the nodes in each element. The gradients are obtained by deriving the element shape functions. However, it has been recently shown by Rodríguez-Galán, Segurado, and Romero (2017) that this simple approach might present convergence problems, and alternative methods to compute the gradients based on recovery techniques are more reliable (Han, Ma, Roters, & Raabe, 2007; Rodríguez-Galán et al., 2017).

3.3.2 Higher Order SGCP

There are several alternative formulations of higher order SGCP models. One of the most extended ones was proposed by Gurtin (2002, 2008) and Gurtin and Needleman (2005). The principal concepts and equations of this theory will be reviewed here because they are common to many other gradient models (Bardella, 2006; Bardella et al., 2013; Borg et al., 2008; Niordson & Kysar, 2014). In this theory, the plastic slip in each system, γ^k, is included in the constitutive equation as a kinematic variable (independent state variable). A central point to the model is that the work associated with each independent kinematic process should be accounted for in the energy balance. Therefore, microforces conjugated with the slip and slip gradients appear in the formulation. In the absence of body forces, the virtual power principle for a domain Ω with boundary $\partial\Omega$ under external macroscopic surface traction $\mathbf{t}$ and a microscopic surface traction Ξ^k for each slip system reads

$$\int_{\partial\Omega} \mathbf{t}\cdot\tilde{\mathbf{v}}dA + \sum_{\alpha}\int_{\partial\Omega}\Xi^k\tilde{\dot{\gamma}}^k dA = \int_{\Omega}\boldsymbol{\sigma}:\mathrm{grad}(\tilde{\mathbf{v}})dV + \sum_{k}\int_{\Omega}\left(\pi^k\tilde{\dot{\gamma}}^k + \boldsymbol{\kappa}^k\cdot\mathrm{grad}\left(\tilde{\dot{\gamma}}^k\right)\right)dV \tag{45}$$

where the left- and right-hand sides of this equation correspond to the external and the internal power, respectively, and the fields $\tilde{\mathbf{v}}$ and $\tilde{\dot{\gamma}}^k$ are the virtual fields of the velocity and slip rate, respectively. π^k and $\boldsymbol{\kappa}^k$ are the higher order stresses conjugated with the slip rate (scalar microforce) and the gradient of the slip rate (vector microforce). From the virtual power expression, the resulting balance equations are the classical macroscopic force and momentum balances and the microforce balance,

$$\text{div}\boldsymbol{\kappa}^k + \tau^k - \pi^k = 0 \quad \text{for} \quad k = 1, .., n. \tag{46}$$

Note that the macroscopic stress tensor enters in the microforce balance through the resolved shear stress τ^k, Eq. (13). In this framework, π^k can be viewed as an internal resistance force to the slip caused by the other dislocations and the microforce vector $\boldsymbol{\kappa}^k$ represents the interaction of dislocations through surfaces.

The second ingredient of the theory is the constitutive equation for the microstresses. In Gurtin (2002), the elastic free energy, Eq. (6), is augmented by a defect energy Ψ that depends on the dislocation tensor $\mathbf{G}$, Eq. (42), defined in Cermelli and Gurtin (2001). If a quadratic expression is chosen, the defect energy Ψ is given by

$$\Psi = \frac{1}{2}\lambda|\mathbf{G}|^2 \tag{47}$$

where λ is an scalar material parameter, with force dimensions, that can be split in a reference modulus of strength π_0 with dimensions of stress and the square of a characteristic length scale l (Bardella, 2006),

$$\lambda = \pi_0 l^2. \tag{48}$$

From the expression of the defect energy, Eq. (47), a linear relation is found between the microforce vector and the dislocation tensor,

$$\boldsymbol{\kappa}^k = \lambda J^{-1}\mathbf{F}^e\left(\mathbf{m}^k \times \mathbf{G}\mathbf{s}^k\right). \tag{49}$$

Finally, higher order boundary conditions should be applied in the external boundaries. Two types of microscopic boundary conditions are usually considered in higher order SGCP, *microfree* and *microhard* boundary conditions. A surface S has microfree boundary conditions if

$$\boldsymbol{\kappa}^k \cdot \mathbf{n} = 0 \text{ on } S, \quad k = 1, 2, \ldots, n \tag{50}$$

and this implies that slip through that surface S is not restricted. Thus, the microforce vector is directly linked to the applied macroscopic traction on that boundary. The microhard boundary condition establishes that the plastic slip on a surface S is restricted,

$$\dot{\gamma}^k = 0 \text{ on } S, \quad k = 1, 2, \ldots, n \tag{51}$$

This condition emulates, e.g., a passivation layer that does not allow the dislocation flux. Other special boundary conditions can be applied at the grain boundaries based on the flow of Burgers vector at and across the boundary surface, as discussed by Gurtin and Needleman (2005).

The finite element implementation of higher order SGCP models is a complicated task. Firstly, it requires higher order continuity associated with higher order strain gradient terms, and this condition is normally addressed by means of mixed finite elements formulations (Shu & Fleck, 1999). Secondly, the application of standard implicit integration algorithms to higher order theories presents difficulties in both efficiency and accuracy for some relevant boundary value problems, as shown in the implementation of Gurtin macroscopic theory of 2005 proposed by Lele and Anand (2008). Different numerical alternatives have been used to improve the efficiency of finite element implementations, using discontinuous Galerkin (Ostien & Garikipati, 2008), explicit integration (Borg et al., 2008), or the recent implicit viscoplastic approach proposed by Panteghihi and Bardella (2016) using a special viscoplastic potential. In addition, the implementation of higher order SGCP models implies the additional computational cost of introducing a very large number of kinematic variables. For instance, the number of kinematic variables per node will increase from 3 to 15 if a theory similar to the one presented by Gurtin (2002) is implemented in a three-dimensional model of a FCC material. Thus, higher order SGCP models have not been applied to simulate the behavior of complex three-dimensional RVEs of polycrystals and most of the computational homogenization studies in the literature using SGCP are based in lower-order theories. An interesting and promising alternative to overcome some of the limitations of the higher order SGCP models is the use of an FFT-based framework to solve the boundary value problem of higher order models, as shown by Lebensohn and Needleman (2016), who used the small-strain version of Gurtin (2002) theory to analyze grain size effects in polycrystals.

3.4 Deformation by Twinning

In addition to dislocation slip, plastic deformation in some metals with low symmetry crystal structures (such as HCP Ti, Mg, and Zr) can occur by twinning (Mahajan & Williams, 1973). This deformation mechanism is present when there are not enough slip systems for an arbitrary shape change of the crystal and twinning provides an additional mechanism to accommodate deformation. A mechanical twin formally corresponds to a sheared volume for which the lattice orientation is transformed into its mirror image across a so-called twin or habit plane (oblique dividing plane defined by the twinning direction). The sheared region of the crystal undergoes an irreversible shear deformation whose value is determined by the lattice geometry and twin plane.

Mechanical twinning is a process that involves two steps. The first one is the nucleation and propagation of a thin twin band across the grain, starting normally from a grain boundary. Afterward, the twinned region propagates in the direction perpendicular to the twin plane and eventually occupies most of the parent grain (Mahajan & Williams, 1973).

The geometrical description of deformation twinning was established in 1965 (Bilby & Crocker, 1965). Since then, many formulations have been published, in which the geometrical description and the driving forces for twinning have been studied for different materials (Christian & Mahajan, 1995). The introduction of deformation by twinning in CP models was taken into account since the early developments of CP (Houtte, 1978; Tomé, Lebensohn, & Kocks, 1991), specially due to its relevance on the deformation of HCP metallic alloys. The mechanical process of twinning is very complex and strong simplifications are needed for the introduction in a CP framework. The first models were applied in the context on mean-field approximations (Houtte, 1978; Tomé et al., 1991) and one of the main challenges of introducing twinning was accounting for the large number of crystal orientations that appear due the formation of twins within each grain. Houtte (1978) tracked the evolution of the volume fraction of the twinned regions in each grain, but reorientation was performed only on selected full grains following a statistical criterion based on the evolution of volume fraction of the twinned regions in the grain and in the entire polycrystal. Thus, the number of grains was kept constant in this approach. An improvement with respect to this approach was proposed by Tomé et al. (1991) leading to the predominant twinning reorientation (PTR) model that has been extensively used within the framework of the VPSC formulation to account for deformation twinning (Proust, Tomé, & Kaschner, 2007). In addition, an alternative model was also proposed in Tomé et al. (1991) based on a discretization of the orientation space $SO3$ and the representation of the orientations as volume fractions in the discrete space. With this framework, the accommodation of deformation by twinning does not increase the number of orientations in the model and nonpredominant orientations can also be represented. Nevertheless, the main limitation of these models is that they were developed within the context of mean-field approximations for polycrystal plasticity and, therefore, cannot be easily included in single crystal plasticity formulations.

The first model that was able to describe twinning deformation of the individual single crystals was developed by Kalidindi (1998), and other similar models have been developed afterward (Abdolvand & Daymond, 2013a, 2013b; Chang & Kochmann, 2015; Kowalczyk-Gajewska, 2010; Mareau & Daymond, 2016; Staroselsky & Anand, 2003). Kalidindi's model has been extensively used (Herrera-Solaz, Hidalgo-Manrique, et al., 2014; Herrera-Solaz, LLorca, Dogan, Karaman, & Segurado, 2014; Hidalgo-Manrique et al., 2015; Jung et al., 2017; Khan, Zahedi, & Siddiqui, 2016; Zhang & Joshi, 2012) and will be briefly described here.

The twinning model of Kalidindi (1998) is a two-scale model. Each material point has a substructure and is divided into two phases, a parent region and a twinned region (Fig. 3), which might be potentially formed by N_{tw} subregions. Each subregion belongs to a given twinning system α and its volume fraction is f^α. Thus, the parent region volume fraction is given by $1 - \sum_{\alpha=1}^{N_{tw}} f^\alpha$. The material point is considered a *composite* material in which the iso-strain hypothesis holds ($\mathbf{F}$ and $\mathbf{F}^e$ are the same in all regions). The plastic deformation is the result of three mechanisms, and the plastic velocity gradient in the intermediate configuration contains three terms, the standard one for dislocation slip, Eq. (11), and two new contributions. The first extra contribution, $\mathbf{L}^p_{tw}$, stands for the rate of deformation due to the twin transformation of a volume fraction df^α of the parent phase

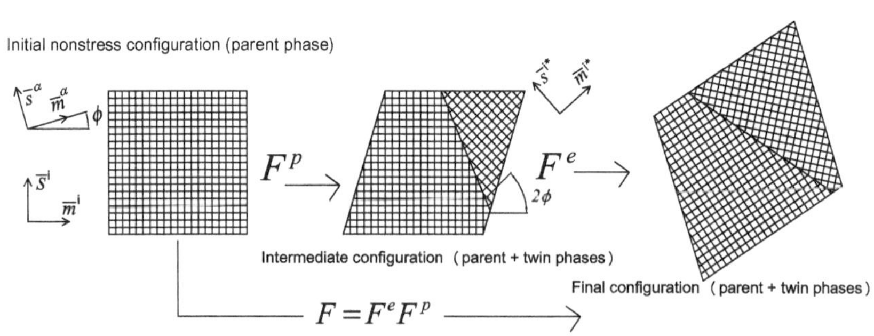

Fig. 3 Multiplicative decomposition indicating material point subdivision in parent and twin phases. *Reprinted from Herrera-Solaz, V., LLorca, J., Dogan, E., Karaman, I., & Segurado, J. (2014). An inverse optimization strategy to determine single crystal mechanical behavior from polycrystal tests: Application to AZ31 Mg alloy. International Journal of Plasticity, 57, 1–15.*

$$\mathbf{L}^p_{tw} = \sum_{\alpha=1}^{N_{tw}} \dot{f}^{\alpha} \gamma_{tw}\, \mathbf{s}^{\alpha}_{tw} \otimes \mathbf{m}^{\alpha}_{tw} \tag{52}$$

where $\dot{f}^{\alpha} = df^{\alpha}/dt$ is the transformation rate in the twin system α, $\mathbf{m}^{\alpha}_{tw}$ and $\mathbf{s}^{\alpha}_{tw}$ are unit vectors along the twin plane normal and the twining direction, respectively, and γ_{tw} stands for the characteristic shear strain of the twinning mode. It is interesting to note that, contrary to dislocation slip, the accumulated plastic deformation by twinning is limited, and the maximum plastic shear that can be by accommodated is γ_{tw} that corresponds to the full transformation of a material point into a given twin variant.

The second contribution to plastic slip corresponds to the slip in the N_{sl-tw} slip systems of the transformed regions, $\mathbf{L}^p_{sl-tr}$, which can be expressed as,

$$\mathbf{L}^p_{sl-tr} = \sum_{\alpha=1}^{N_{tw}} f^{\alpha} \left(\sum_{k^*=1}^{N_{sl-tw}} \dot{\gamma}^{k^*} \mathbf{s}^{k^*}_{sl} \otimes \mathbf{m}^{k^*}_{sl} \right) \tag{53}$$

where $\mathbf{s}^{k^*}_{sl}$ and $\mathbf{m}^{k*}_{sl}$ stand for the unit vectors in the slip and normal directions to the slip plane. They can be computed from the orientation of the slip planes k in the parent grain by means of the matrix, $\mathbf{Q}^{\alpha}$,

$$\mathbf{Q}^{\alpha} = 2\mathbf{m}^{\alpha}_{tw} \otimes \mathbf{m}^{\alpha}_{tw} - \mathbf{I}. \tag{54}$$

that takes into account the rotation of the crystal due to twinning in the α plane.

An evolution equation defining $\dot{f}^{\alpha}$, Eq. (52), has to be specified to complete the flow rule for twinning. It is well accepted that the driving force for twin growth is the resolved shear stress on the twinning system. Accordingly, a viscous law depending on the resolved shear stress was proposed in Kalidindi (1998), which is equivalent to the expression used for slip in Asaro and Needleman (1985). Thus,

$$\dot{f}^{\alpha} = \dot{f}_0 \left(\frac{\langle \tau^{\alpha} \rangle}{g^{\alpha}} \right)^{\frac{1}{m}} \quad \text{with} \quad \langle \tau \rangle = \begin{cases} \tau & \text{if } \tau \geq 0 \\ 0 & \text{if } \tau < 0 \end{cases} \tag{55}$$

where $\dot{f}_0$ is a reference transformation rate, m the strain rate-sensitivity parameter and $\langle x \rangle$ stand for the Macaulay brackets that are introduced to account for the polar nature of twinning deformation. τ^{α} is the resolved stress in the twinning system α and its value is given by

$$\tau^{\alpha} = \mathbf{S}^{parent} : \mathbf{s}^{\alpha}_{tw} \otimes \mathbf{m}^{\alpha}_{tw} \tag{56}$$

where $\mathbf{S}^{parent}$ denotes the value of the second Piola–Kirchhoff stress in the parent region. Note that the stresses in the parent and twinned region are different due to the iso-strain approach. In particular, $\mathbf{S}^{parent}$ is obtained directly from the linear relation between the elastic strain in the intermediate configuration, $\mathbf{E}^e$ (common for both parent and twin phases)

$$\mathbf{S}^{parent} = \mathbb{L}\mathbf{E}^e \tag{57}$$

where $\mathbb{L}$ stands for the fourth-rank elastic stiffness tensor of the crystal in its original orientation. The Piola–Kirchhoff stress tensor for the full integration point (containing the parent and all the twinned phases) in the intermediate configuration, $\mathbf{S}$, is obtained in this case from the volume-averaged stress tensors in the different phases

$$\mathbf{S} = \left(1 - \sum_{\alpha=1}^{N_{tw}} f^\alpha\right)\mathbf{S}^{parent} + \sum_{\alpha=1}^{N_{tw}} f^\alpha \mathbf{S}^\alpha \tag{58}$$

where $\mathbf{S}^\alpha$ is the stress in each of the twinned regions, and is obtained from an expression equivalent to Eq. (57).

4. VISCOPLASTIC SELF-CONSISTENT HOMOGENIZATION OF POLYCRYSTALS

The computation of the effective mechanical response and texture evolution of polycrystalline materials using homogenization theory has a long tradition (Sachs, 1928; Taylor, 1938) and self-consistent approximations have been extensively used to deal with this problem. The 1-site VPSC theory of polycrystal deformation can be traced back to the work of Molinari et al. (1987a), who established a homogenization procedure based on an iterative method involving the computation of integrals in ellipsoidal domains of the infinite-medium Green's function, customarily used in the solution of the PDEs governing the micromechanical response of heterogeneous materials. This formulation was implemented numerically by Lebensohn and Tomé (1993) taking into account the polycrystal anisotropy, leading to the first version of the VPSC code. Since its inception, the VPSC code has experienced several improvements and extensions, e.g., recrystallization (Wenk, Canova, Brechet, & Flandin, 1997); 2-site approximation for 2-phase polycrystals (Lebensohn & Canova, 1997); VPSC modeling of lamellar structures (Lebensohn, Uhlenhut, Hartig, & Mecking, 1998); relative directional compliance interaction (Tomé, 1999); second-order linearization (Lebensohn et al., 2007); improved VPSC modeling of twinning using the PTR approach (Proust et al., 2007); dislocation density-based hardening models (Beyerlein & Tomé, 2008); climb and glide VPSC model (Lebensohn, Hartley, Tomé, & Castelnau, 2010); dilatational VPSC for porous polycrystals (Lebensohn, Idiart, Ponte Castañeda, & Vincent, 2011); lattice rotation rate fluctuation calculation (Lebensohn, Zecevic, Knezevic, & McCabe, 2016); improved VPSC for arbitrarily low rate sensitivities (Knezevic, Zecevic, Beyerlein, & Lebensohn, 2016); improved hardening laws for strain-path changes (Wen et al., 2016); VPSC prediction of intragranular misorientation evolution (Zecevic, Pantleon, Lebensohn, McCabe, & Knezevic, 2017), etc. The VPSC homogenization strategy is nowadays extensively used to simulate plastic deformation of polycrystalline aggregates and for interpretation of experimental results in metals, minerals and polymers.

The self-consistent theory is one of the most commonly used homogenization methods to estimate the mechanical response behavior of polycrystals and was originally proposed by Hershley (1954) for linear elastic materials. For nonlinear aggregates (as those formed by grains deforming in the viscoplastic regime), several self-consistent (SC) approximations have been proposed. They differ in the procedure used to linearize the nonlinear mechanical response at the grain level, but all of them eventually make use of the original linear SC theory. They include the secant (Hill, 1965b; Hutchinson, 1976), the tangent (Lebensohn & Tomé, 1993; Molinari, Canova, & Ahzi, 1987b), and the affine (Masson, Bornert, Suquet, & Zaoui, 2000) first-order approximations, which are based on linearization schemes that use the information on field averages at the grain level, and disregard higher order statistical information inside the grains. However, the above assumption may be questionable in materials which present strong directionality and/or large variations in local properties. This is the case of low rate-sensitivity materials, aggregates made of highly anisotropic grains, voided and/or multiphase polycrystals. In all these cases, strong deformation gradients are likely to develop inside grains because of differences in properties with neighbor crystals or phases, including voids.

More accurate nonlinear homogenization methods were developed, mainly due to the work of Ponte Castañeda and collaborators, to overcome the above limitations. These methods use linearization schemes at grain level that also incorporate accessible information on the second moments of the stress field distributions in the grains. These more elaborate SC formulations are based on the concept of a linear comparison material, which expresses the effective potential of the nonlinear viscoplastic polycrystal in terms of that of a linearly viscous aggregate whose properties are determined from suitably designed variational principles. Ponte Castañeda's first variational method was originally proposed for nonlinear composites (Ponte Castañeda, 1991) and then extended to viscoplastic polycrystals (de Botton & Ponte Castañeda, 1995). It makes use of the SC approximation for linearly viscous polycrystals to obtain bounds and estimates for nonlinear viscoplastic polycrystals. The more recent second-order method, proposed for nonlinear composites (Ponte Castañeda, 2002), and later extended to viscoplastic polycrystals (Liu & Ponte Castañeda, 2004), uses the SC approximation for a more general class of linearly viscous polycrystals, having a nonvanishing strain rate at zero stress, to generate even more accurate SC estimates for viscoplastic polycrystals. The implementation of a fully anisotropic second-order approach inside the VPSC code (Lebensohn et al., 2007) has been a necessary step toward improving its predictive capability for polycrystalline materials that exhibit high contrast in local properties. Unavoidably, this improved capability came at the expense of more complex and numerically demanding algorithms. In what follows, the VPSC formulation is first presented using the affine linearization scheme (Masson et al., 2000), and the second-order linearization procedure (Liu & Ponte Castañeda, 2004) is described next.

The self-consistent formulation represents the polycrystal by means of weighted, ellipsoidal, statistically representative (SR) grains. Each of these SR grains represent the average behavior of all the grains with a particular crystallographic orientation and morphology but different local environments. These SR grains should be regarded as representing the behavior of mechanical phases, i.e., all the single crystals with a given orientation (*r*) belong to mechanical phase (*r*) and are represented by the SR grain (*r*). Note the difference between *mechanical phases*, which differ from each other only in terms of crystallographic orientation and/or morphology, and actual *phases*, which differ from each other in crystallographic structure and/or composition. In what follows, *SR grain* (*r*) and *mechanical phase* (*r*) will be used interchangeably. The weights represent volume fractions and are chosen to reproduce the initial texture of the material. Each representative grain is treated as an ellipsoidal viscoplastic inclusion embedded in an effective viscoplastic medium. Both inclusion and medium are anisotropic. Plastic deformation in the inclusion is accommodated by dislocation slip activated by a resolved shear stress. As a consequence of all the above assumptions, the representation of the polycrystalline aggregate under the SC model is nonspace-resolved, and corresponds to an entire class of polycrystals with microstructures consistent with a given statistical distribution.

In general, homogenization models for viscoplastic deformation assumes that the elastic strains are much smaller than the plastic ones and, thus, are neglected. The plastic deformation rates are constitutively related to stress in the current configuration using small-strain kinematics, resulting in relations between Cauchy stress and velocity gradient (instead of the second Piola–Kirchhoff stress and the deformation gradient). Once the velocity gradient is obtained, the evolution of microstructure and micromechanical variables can be calculated by integrating the velocity gradient field, or local (grain) averages of the latter, in small time increments to update the current configuration of the material.

4.1 Local Constitutive Behavior and Homogenization

Assuming small-strain kinematics for the deformation rates in the current configuration, the macroscopic velocity gradient $V_{i,j}$ applied to a polycrystalline aggregate can be decomposed into an average symmetric strain rate $\dot{E}_{ij} = \frac{1}{2}\left(V_{i,j} + V_{j,i}\right)$ and an average antisymmetric rotation rate $\dot{\Omega}_{ij} = \frac{1}{2}\left(V_{i,j} - V_{j,i}\right)$. The plastic component of the deformation is assumed to be much larger than the elastic part and, therefore, the flow is incompressible. The viscoplastic constitutive behavior at each material point $\mathbf{x}$ is described by means of a nonlinear rate-sensitive equation as

$$\dot{\boldsymbol{\epsilon}}(\mathbf{x}) = \sum_{k=1}^{N_k} \mathbf{Z}_s^k(\mathbf{x})\dot{\gamma}^k(\mathbf{x}) \tag{59}$$

where $\dot{\boldsymbol{\epsilon}}$ is the local strain rate and $\mathbf{Z}_s^k(\mathbf{x}) = \frac{1}{2}(\mathbf{s}^k(\mathbf{x}) \otimes \mathbf{m}^k(\mathbf{x}) + \mathbf{m}^k(\mathbf{x}) \otimes \mathbf{s}^k(\mathbf{x}))$ is the symmetric Schmid tensor associated with slip system k and $\dot{\gamma}^k$ stands for the local shear rate on slip system k. The shear rate on each system follows a power-law relation (Peirce, 1983)

$$\begin{aligned} \dot{\gamma}^k(\mathbf{x}) &= \dot{\gamma}_0 \left(\frac{\left|\tau^k(\mathbf{x})\right|}{g^k(\mathbf{x})} \right)^n \operatorname{sign}\left(\tau^k(\mathbf{x})\right) \\ &= \dot{\gamma}_0 \left(\frac{\left|\mathbf{Z}_s^k(\mathbf{x}) : \boldsymbol{\sigma}'(\mathbf{x})\right|}{g^k(\mathbf{x})} \right)^n \operatorname{sign}\left(\mathbf{Z}_s^k(\mathbf{x}) : \boldsymbol{\sigma}'(\mathbf{x})\right) \end{aligned} \tag{60}$$

where g^k is the CRSS of slip system k, $\dot{\gamma}_0$ is the reference strain rate, n the rate-sensitivity exponent and $\dot{\boldsymbol{\epsilon}}$ and $\boldsymbol{\sigma}'$ stand for the strain rate and the deviatoric part of the Cauchy stress. If the shear rates are known, the lattice rotation rate, $\dot{\omega}^p$ (or plastic spin) associated with slip activity at a single crystal material point $\mathbf{x}$ is given by:

$$\dot{\boldsymbol{\omega}}^{\mathbf{p}}(\mathbf{x}) = \sum_k \mathbf{Z}_a^k(\mathbf{x})\dot{\gamma}^k(\mathbf{x}) \tag{61}$$

where $\mathbf{Z}_a^k(\mathbf{x}) = \frac{1}{2}(\mathbf{s}^k(\mathbf{x}) \otimes \mathbf{m}^k(\mathbf{x}) - \mathbf{m}^k(\mathbf{x}) \otimes \mathbf{s}^k(\mathbf{x}))$ is the antisymmetric Schmid tensor. Note that although Eqs. (59)–(61) can be used to deal with crystal deformation by slip and twinning, only slip will be considered in the examples that follow in the context of both mean-field and full-field approaches, to avoid the additional complication of twinning reorientation. Moreover, the constitutive behavior described by Eqs. (59)–(61) does not consider other possible crystal deformation mechanisms and microstructural evolution processes, such as climb, grain boundary sliding or recrystallization.

Assuming that the following linear relations, i.e., approximations of the actual nonlinear relations, Eqs. (59) and (60), hold for the SR grain (r), it can be written

$$\dot{\boldsymbol{\epsilon}}(\mathbf{x}) = \mathbb{M}^{(r)}\boldsymbol{\sigma}'(\mathbf{x}) + \dot{\boldsymbol{\epsilon}}^{0(r)} \tag{62}$$

where $\mathbb{M}^{(r)}$ and $\dot{\boldsymbol{\epsilon}}^{0(r)}$ are the viscous compliance tensor and back-extrapolated strain rate (strain rate under zero stress) of grain (r), respectively, which depend on the linearization assumption. For example, under the affine linearization, they are given by

$$\mathbb{M}^{(r)} = n\dot{\gamma}_0 \sum_{k=1}^{N_k} \frac{\mathbf{Z}_s^{k(r)} \otimes \mathbf{Z}_s^{k(r)}}{g^{k(r)}} \left(\frac{|\mathbf{Z}_s^{k(r)} : \boldsymbol{\sigma}'^{(r)}|}{g^{k(r)}} \right)^{n-1} \tag{63}$$

$$\dot{\boldsymbol{\epsilon}}^{0(r)} = (1-n)\,\dot{\gamma}_0 \sum_{k=1}^{N_k} \left(\frac{\left|\mathbf{Z}_s^{k(r)} : \boldsymbol{\sigma}'^{(r)}\right|}{g^{k(r)}} \right)^n \operatorname{sign}\left(\mathbf{Z}_s^{k(r)} : \boldsymbol{\sigma}'^{(r)}\right) \tag{64}$$

where the index (r) on the magnitudes on the right-hand side of these equations indicates average over the SR (r), e.g., $\boldsymbol{\sigma}'^{(r)} = \langle \boldsymbol{\sigma}'(\mathbf{x}) \rangle^{(r)} = \langle \boldsymbol{\sigma}' \rangle^{(r)}$.

Following Hill (1965b) and Hutchinson (1976), the homogenized behavior of a linear heterogeneous medium whose local behavior is described by Eq. (62) can also be described by an analogous linear relation at the effective (macroscopic) level

$$\dot{\mathbf{E}} = \overline{\mathbb{M}}\boldsymbol{\Sigma}' + \dot{\mathbf{E}}^0 \tag{65}$$

where $\dot{\mathbf{E}}$ and $\boldsymbol{\Sigma}'$ are the overall (macroscopic) deviatoric strain rate and stress tensor, respectively, and $\overline{\mathbb{M}}$ and $\dot{\mathbf{E}}^0$ stand for the viscous compliance tensor and the back-extrapolated strain rate, respectively, of an a priori unknown homogeneous equivalent medium (HEM). The response of this HEM is obtained by the linear SC method. The problem underlying the SC method is that of an inhomogeneous domain (*r*) of moduli $\mathbb{M}^{(r)}$ and $\dot{\boldsymbol{\epsilon}}^{0(r)}$, embedded in an infinite medium of moduli $\overline{\mathbb{M}}$ and $\dot{\mathbf{E}}^0$. Invoking the concept of the equivalent inclusion (Mura, 1987), the local constitutive behavior in grain (*r*) can be rewritten as

$$\dot{\boldsymbol{\epsilon}}(\mathbf{x}) = \overline{\mathbb{M}}\boldsymbol{\sigma}'(\mathbf{x}) + \dot{\mathbf{E}}^0 + \dot{\boldsymbol{\epsilon}}^*(\mathbf{x}) \tag{66}$$

where $\dot{\boldsymbol{\epsilon}}^*(\mathbf{x})$ is an eigenstrain rate field, which follows from replacing the inhomogeneity by an equivalent inclusion. Rearranging and subtracting Eq. (65) from Eq. (66) gives

$$\tilde{\boldsymbol{\sigma}}'(\mathbf{x}) = \overline{\mathbb{L}}\left[\tilde{\dot{\boldsymbol{\epsilon}}}(\mathbf{x}) - \dot{\boldsymbol{\epsilon}}^*(\mathbf{x})\right] \tag{67}$$

where the symbol "$\sim$" denotes local deviations from macroscopic values of the corresponding magnitudes, and $\overline{\mathbb{L}} = \overline{\mathbb{M}}^{-1}$. Combining Eq. (67) with the equilibrium condition gives

$$\sigma_{ij,j}(\mathbf{x}) = \tilde{\sigma}_{ij,j}(\mathbf{x}) = \tilde{\sigma}'_{ij,j}(\mathbf{x}) + \tilde{\sigma}^m_{,i}(\mathbf{x}) \tag{68}$$

where σ_{ij} and σ^m are the Cauchy stress tensor and Cauchy the mean stress, respectively. From the relation between the strain rate and the velocity gradient deviations, $\tilde{\dot{\epsilon}}_{ij}(\mathbf{x}) = \frac{1}{2}\left(\tilde{v}_{i,j}(\mathbf{x}) + \tilde{v}_{j,i}(\mathbf{x})\right)$, and taking into account the incompressibility condition associated with plastic deformation,

$$\left|\begin{array}{l} \overline{L}_{ijkl}\,\tilde{v}_{k,lj}(\mathbf{x}) + \tilde{\sigma}^m_{,i}(\mathbf{x}) + \varphi_{ij,j}(\mathbf{x}) = 0 \\ \tilde{v}_{k,k}(\mathbf{x}) = 0 \end{array}\right. \tag{69}$$

where

$$\varphi_{ij}(\mathbf{x}) = -\overline{L}_{ijkl}\dot{\epsilon}^*_{kl}(\mathbf{x}) \tag{70}$$

is a heterogeneity or *polarization field*, and its divergence $f_i(\mathbf{x}) = \varphi_{ij,j}(\mathbf{x})$ is an artificial external force field applied to the material.

The set of Eq. (69) consists of four differential equations with four unknowns: three are the components of the deviation from the average value of the velocity vector $\tilde{v}_i(\mathbf{x})$, and one is the deviation of the mean stress

$\tilde{\sigma}^m(\mathbf{x})$. A set of N linear differential equations with N unknown functions and a polarization term can be solved using the Green function method. $G_{km}(\mathbf{x})$ and $H_m(\mathbf{x})$ are the Green functions associated with $\tilde{v}_i(\mathbf{x})$ and $\tilde{\boldsymbol{\sigma}}^m(\mathbf{x})$, respectively, which solve the auxiliary problem of a unitary volumetric force with a single nonvanishing m-component,

$$\left| \begin{array}{l} \overline{L}_{ijkl}\, G_{km,lj}\,(\mathbf{x}-\mathbf{x}') + H_{m,i}\,(\mathbf{x}-\mathbf{x}') + \delta_{im}\,\delta\,(\mathbf{x}-\mathbf{x}') = 0 \\ G_{km,k}(\mathbf{x}-\mathbf{x}') = 0 \end{array} \right. \tag{71}$$

where the Green's function $G_{km}(\mathbf{x}-\mathbf{x}')$ is the deviation in velocity component in the x_k-direction at point $\mathbf{x}$ when a unit body force in the x_m-direction is applied at point $\mathbf{x}'$ in an infinitely extended material and $H_m(\mathbf{x}-\mathbf{x}')$ is the corresponding response in mean stress.

Once the solution of Eq. (71) is obtained, the velocity field is given by the convolution integral:

$$\tilde{v}_k(\mathbf{x}) = \int_{\mathbb{R}^3} G_{ki}(\mathbf{x}-\mathbf{x}')f_i(\mathbf{x}')d\mathbf{x}' \tag{72}$$

The set of Eq. (71) can be solved using Fourier transforms. Expressing the Green functions in terms of their inverse Fourier transforms, the differential equations (71) can be transformed into an algebraic set

$$\left| \begin{array}{l} \alpha_j\,\alpha_l\,\overline{L}_{ijkl}\,k^2\,\hat{G}_{km}\,(\boldsymbol{\xi}) + \alpha_i\,ik\hat{H}_m\,(\boldsymbol{\xi}) = \delta_{im} \\ \alpha_k\,k^2\,\hat{G}_{km}(\boldsymbol{\xi}) = 0 \end{array} \right. \tag{73}$$

where k and $\boldsymbol{\alpha}$ are the modulus and the unit vector associated with a point of Fourier space $\boldsymbol{\xi} = k\boldsymbol{\alpha}$, respectively. Calling $A'_{ik}(\boldsymbol{\alpha}) = \alpha_j\,\alpha_l\,\overline{L}_{ijkl}$, the set (73) can be expressed as a matrix product $\mathbf{AB} = \mathbf{C}$ where $\mathbf{A}$, $\mathbf{B}$, and $\mathbf{C}$ are given by

$$\mathbf{A}(\boldsymbol{\alpha}) = \begin{bmatrix} A'_{11} & A'_{12} & A'_{13} & \alpha_1 \\ A'_{21} & A'_{22} & A'_{23} & \alpha_2 \\ A'_{31} & A'_{32} & A'_{33} & \alpha_3 \\ \alpha_1 & \alpha_2 & \alpha_3 & 0 \end{bmatrix}$$

$$\mathbf{B} = \begin{bmatrix} k^2\hat{G}_{11} & k^2\hat{G}_{12} & k^2\hat{G}_{13} \\ k^2\hat{G}_{21} & k^2\hat{G}_{22} & k^2\hat{G}_{23} \\ k^2\hat{G}_{31} & k^2\hat{G}_{32} & k^2\hat{G}_{33} \\ ik\hat{H}_1 & ik\hat{H}_2 & ik\hat{H}_3 \end{bmatrix} \qquad \mathbf{C} = \begin{bmatrix} 1 & 0 & 0 \\ 0 & 1 & 0 \\ 0 & 0 & 1 \\ 0 & 0 & 0 \end{bmatrix} \tag{74}$$

Using the explicit form of matrix $\mathbf{C}$,

$$\mathbf{B}=\mathbf{A}^{-1}\mathbf{C}=\begin{bmatrix} A_{11}^{-1} & A_{12}^{-1} & A_{13}^{-1} \\ A_{21}^{-1} & A_{22}^{-1} & A_{23}^{-1} \\ A_{31}^{-1} & A_{32}^{-1} & A_{33}^{-1} \\ A_{41}^{-1} & A_{42}^{-1} & A_{43}^{-1} \end{bmatrix}. \tag{75}$$

gives

$$k^2\hat{G}_{ij}(\boldsymbol{\xi})=A_{ij}^{-1}(\boldsymbol{\alpha}) \quad (i,j=1,3) \tag{76}$$

Comparing Eqs. (74) and (75), since the components of $\mathbf{A}$ are real functions of $\boldsymbol{\alpha}$, so are those of $k^2\hat{G}_{ij}(\boldsymbol{\xi})$. This property leads to real integrals in the derivation that follows.

Knowing the Green's function expression in Fourier space, the solution of the eigenstrain rate problem can be obtained using convolution integrals. Taking partial derivatives to Eq. (72) leads to

$$\tilde{v}_{k,l}(\mathbf{x})=\int_{\mathbb{R}^3} G_{ki,l}(\mathbf{x}-\mathbf{x}')f_i(\mathbf{x}')\,d\mathbf{x}'. \tag{77}$$

Replacing the expression of the artificial volumetric force field in Eq. (77), recalling that $\partial G_{ij}(\mathbf{x}-\mathbf{x}')/\partial\mathbf{x}=-\partial G_{ij}(\mathbf{x}-\mathbf{x}')/\partial\mathbf{x}'$, integrating by parts, and using the divergence theorem (Mura, 1987),

$$\tilde{v}_{k,l}(\mathbf{x})=\int_{\mathbb{R}^3} G_{ki,jl}(\mathbf{x}-\mathbf{x}')\,\varphi_{ij}(\mathbf{x}')\,d\mathbf{x}'. \tag{78}$$

The integral equation (78) provides an exact implicit solution to the problem. Furthermore, it is known from Eshelby's elastic inclusion formalism that the stress and strain are constant over the domain of the inclusion (r), Ω_r, if the eigenstrain is uniform over an ellipsoidal domain where the stiffness tensor is uniform. This suggests the use of an a priori unknown constant polarization within the volume of the ellipsoidal inclusion and allows to average the local field of Eq. (78) over the domain and obtain an average strain rate inside the inclusion of the form

$$\tilde{v}_{k,l}^{(r)}=\left(-\frac{1}{\Omega_r}\int_{\Omega_r}\int_{\Omega_r} G_{ki,jl}(\mathbf{x}-\mathbf{x}')\,d\mathbf{x}\,d\mathbf{x}'\right)\overline{L}_{ijmn}\dot{\epsilon}_{mn}^{*(r)} \tag{79}$$

where $\tilde{v}_{k,l}^{(r)}$ and $\dot{\epsilon}_{mn}^{*(r)}$ have to be interpreted as average quantities inside the inclusion. Expressing the Green's function in terms of the inverse Fourier transform and taking derivatives,

$$\tilde{\nu}_{k,l}^{(r)} = \left(\frac{1}{8\pi^3\Omega_r} \int_{\Omega_r} \int_{\Omega_r} \int_{\mathbb{R}^3} \alpha_j \alpha_l \left(k^2 \hat{G}_{ki}(\boldsymbol{\xi}) \right) \exp\left[-i\boldsymbol{\xi}(\mathbf{x} - \mathbf{x}')\right] d\boldsymbol{\xi} d\mathbf{x}\, d\mathbf{x}' \right)$$
$$\overline{L}_{ijmn} \dot{\epsilon}_{mn}^{*(r)} = T_{klij} \overline{L}_{ijmn} \dot{\epsilon}_{mn}^{*(r)}. \tag{80}$$

Writing $d\boldsymbol{\xi}$ in spherical coordinates ($d\boldsymbol{\xi} = k^2 \sin\theta\, dk\, d\theta\, d\varphi$) and using the relation (76), the Green's interaction tensor T_{klij} can be expressed as

$$T_{klij} = \frac{1}{8\pi^3\Omega_r} \int_0^{2\pi} \int_0^{\pi} \alpha_j\, \alpha_l\, A_{ki}^{-1}(\boldsymbol{\alpha})\, \Lambda(\boldsymbol{\alpha})\, \sin\theta\, d\theta\, d\varphi \tag{81}$$

where

$$\Lambda(\boldsymbol{\alpha}) = \int_0^{\infty} \left(\int_{\Omega_r} \int_{\Omega_r} \exp\left[-i\boldsymbol{\xi}(\mathbf{x} - \mathbf{x}')\right] d\mathbf{x}\, d\mathbf{x}' \right) k^2 dk. \tag{82}$$

Integrating Eq. (82) inside an ellipsoidal grain of radii (a, b, c) and replacing in Eq. (81) leads to

$$T_{klij} = \frac{abc}{4\pi} \int_0^{2\pi} \int_0^{\pi} \frac{\alpha_j\, \alpha_l\, A_{ki}^{-1}(\boldsymbol{\alpha})}{[\rho(\boldsymbol{\alpha})]^3} \sin\,\theta\, d\theta\, d\varphi \tag{83}$$

where $\rho(\boldsymbol{\alpha}) = \left[(a\alpha_1)^2 + (b\alpha_2)^2 + (c\alpha_3)^2\right]^{1/2}$. The symmetric and antisymmetric Eshelby tensors (functions of and the shape of the ellipsoidal inclusion, representing the morphology of the SR grains) are defined as

$$S_{ijkl} = \frac{1}{4}\left(T_{ijmn} + T_{jimn} + T_{ijnm} + T_{jinm} \right) \overline{L}_{mnkl} \tag{84}$$

$$P_{ijkl} = \frac{1}{4}\left(T_{ijmn} - T_{jimn} + T_{ijnm} - T_{jinm} \right) \overline{L}_{mnkl} \tag{85}$$

and the average strain rate and rotation rate deviations, $\tilde{\dot{\boldsymbol{\epsilon}}}^{(r)}$ and $\tilde{\dot{\boldsymbol{\omega}}}^{(r)}$, in the ellipsoidal domain are obtained by taking symmetric and antisymmetric components to Eq. (80) and using Eqs. (84)–(85)

$$\tilde{\dot{\boldsymbol{\epsilon}}}^{(r)} = \mathbb{S}\dot{\boldsymbol{\epsilon}}^{*(r)} \tag{86}$$

$$\tilde{\dot{\boldsymbol{\omega}}}^{(r)} = \mathbb{P}\dot{\boldsymbol{\epsilon}}^{*(r)} = \mathbb{P}\mathbb{S}^{-1}\tilde{\dot{\boldsymbol{\epsilon}}}^{(r)} \tag{87}$$

where $\tilde{\dot{\boldsymbol{\epsilon}}}^{(r)} = \dot{\mathbf{E}} - \dot{\boldsymbol{\epsilon}}^{(r)}$ and $\tilde{\dot{\boldsymbol{\omega}}}^{(r)} = \dot{\boldsymbol{\Omega}} - \dot{\boldsymbol{\omega}}^{(r)}$ are deviations of the average strain rate and rotation rate inside the inclusion, respectively, with respect to the corresponding overall magnitudes, and $\dot{\boldsymbol{\epsilon}}^{*(r)}$ is the average eigenstrain rate in the inclusion. Therefore, the lattice rotation rate field is given by

$$\dot{\boldsymbol{\omega}}(\mathbf{x}) = \dot{\boldsymbol{\Omega}} + \tilde{\dot{\boldsymbol{\omega}}}^{(r)} - \dot{\boldsymbol{\omega}}^{p}(\mathbf{x}) \tag{88}$$

4.2 Interaction and Localization Equations

Taking volume averages over the domain of the inclusion on both sides of Eq. (67) gives

$$\tilde{\boldsymbol{\sigma}}'^{(r)} = \mathbb{L}\left(\tilde{\dot{\boldsymbol{\epsilon}}}^{(r)} - \dot{\boldsymbol{\epsilon}}^{*(r)}\right) \tag{89}$$

and replacing the eigenstrain rate given by Eq. (86) into Eq. (89), leads to the *interaction equation*:

$$\tilde{\dot{\boldsymbol{\epsilon}}}^{(r)} = -\tilde{\mathbb{M}}\tilde{\boldsymbol{\Sigma}}'^{(r)} \tag{90}$$

where the interaction tensor is given by:

$$\tilde{\mathbb{M}} = (\mathbb{I} - \mathbb{S})^{-1}\mathbb{S}\overline{\mathbb{M}}. \tag{91}$$

Replacing the constitutive relations of the inclusion and the effective medium in the interaction equation and, after some manipulation, leads to the *localization equation*:

$$\boldsymbol{\sigma}'^{(r)} = \mathbb{B}^{(r)}\boldsymbol{\Sigma}' + \mathbf{b}^{(r)} \tag{92}$$

where the localization tensors are defined as:

$$\mathbb{B}^{(r)} = \left(\mathbb{M}^{(r)} + \tilde{\mathbb{M}}\right)^{-1}\left(\overline{\mathbb{M}} + \tilde{\mathbb{M}}\right) \tag{93}$$

$$\mathbf{b}^{(r)} = \left(\mathbb{M}^{(r)} + \tilde{\mathbb{M}}\right)^{-1}\left(\dot{\mathbf{E}}^{0} - \dot{\boldsymbol{\epsilon}}^{0(r)}\right) \tag{94}$$

4.3 Self-Consistent Equations

The derivation presented in the previous sections solves the problem of an equivalent inclusion embedded in an effective medium. This result is used in this section to construct a polycrystal model, in which each SR grain (r) stands for an inclusion embedded in an effective medium that represents the polycrystal. The properties of such medium are not known a priori and have to be found through an iterative procedure. If the stress localization equation, Eq. (92) is introduced in the local constitutive equation, Eq. (62), averaged over the SR grain (r),

$$\dot{\boldsymbol{\epsilon}}^{(r)} = \mathbb{M}^{(r)}\mathbb{B}^{(r)}\boldsymbol{\Sigma} + \mathbb{M}^{(r)}\mathbf{b}^{(r)} + \dot{\boldsymbol{\epsilon}}^{0(r)}. \tag{95}$$

Taking volumetric average to Eq. (95) and enforcing the condition that the average of the strain rates over the aggregate has to coincide with the macroscopic quantities, leads to,

$$\dot{\mathbf{E}} = \left\langle \dot{\boldsymbol{\epsilon}}^{(r)} \right\rangle \tag{96}$$

where the brackets denote average over the SR grains weighted by the associated volume fraction. From the macroscopic constitutive relation, Eq. (65), the following *self-consistent equations* are obtained for the HEM's compliance and back-extrapolated term:

$$\overline{\mathbb{M}} = \left\langle \mathbb{M}^{(r)} \mathbb{B}^{(r)} \right\rangle \tag{97}$$

$$\dot{\mathbf{E}}^0 = \left\langle \mathbb{M}^{(r)} \mathbf{b}^{(r)} + \dot{\boldsymbol{\epsilon}}^{0(r)} \right\rangle \tag{98}$$

4.4 Linearization Assumptions

As indicated above, different choices are possible for the linearized behavior at the grain level, and the results of the homogenization scheme depend on this choice. Several first-order linearization schemes are presented below, defined in terms of the stress first-order moment (average) inside SR grain (r).

The *secant approximation* (Hill, 1965b; Hutchinson, 1976) assumes the following linearized secant moduli:

$$\mathbb{M}^{(r)}_{\text{sec}} = \dot{\gamma}_0 \sum_k \frac{\mathbf{Z}_s^{k(r)} \otimes \mathbf{Z}_s^{k(r)}}{g^{k(r)}} \left(\frac{\left| \mathbf{Z}_s^{k(r)} : \boldsymbol{\sigma}'^{(r)} \right|}{g^{k(r)}} \right)^{n-1} \tag{99}$$

$$\dot{\boldsymbol{\epsilon}}^{0(r)}_{\text{sec}} = 0 \tag{100}$$

where the index (r) in $\mathbf{Z}_s^{k(r)}$ and $g^{k(r)}$ indicates uniform (average) values of these magnitudes, corresponding to a given orientation and hardening state associated with the SR grain (r).

Under the *affine approximation* (Masson et al., 2000), the affine moduli are given by

$$\mathbb{M}^{(r)}_{aff} = n\dot{\gamma}_0 \sum_k \frac{\mathbf{Z}_s^{k(r)} \otimes \mathbf{Z}_s^{k(r)}}{g^{k(r)}} \left(\frac{\left| \mathbf{Z}_s^{k(r)} : \boldsymbol{\sigma}'^{(r)} \right|}{g^{k(r)}} \right)^{n-1} \tag{101}$$

$$\dot{\boldsymbol{\epsilon}}^{0(r)}_{aff} = (1-n)\dot{\gamma}_0 \sum_k \left(\frac{\left| \mathbf{Z}_s^{k(r)} : \boldsymbol{\sigma}'^{(r)} \right|}{g^{k(r)}} \right)^{n} \text{sign}\left(\mathbf{Z}_s^{k(r)} : \boldsymbol{\sigma}'^{(r)} \right) \tag{102}$$

In the case of the *tangent approximation* (Lebensohn & Tomé, 1993; Molinari et al., 1987b), the tangent moduli are formally the same as in the affine case: $\mathbb{M}_{tg}^{(r)} = \mathbb{M}_{aff}^{(r)}$ and $\dot{\boldsymbol{\epsilon}}_{tg}^{0(r)} = \dot{\boldsymbol{\epsilon}}_{aff}^{0(r)}$. However, Molinari et al. (1987b) used the secant compliance, Eq. (99), instead of the affine one in order to avoid the iterative adjustment of the macroscopic back-extrapolated term to adjust $\overline{\mathbb{M}}$ (to be denoted $\overline{\mathbb{M}}_{\text{sec}}$), in combination with the tangent-secant relation $\overline{\mathbb{M}}_{tg} = n\overline{\mathbb{M}}_{\text{sec}}$ (Hutchinson, 1976). Then, the expression of the interaction tensor is given by

$$\tilde{\mathbb{M}} = (\mathbb{I} - \mathbb{S})^{-1}\mathbb{S}\overline{\mathbb{M}}_{tg} = n(\mathbb{I} - \mathbb{S})^{-1}\mathbb{S}\overline{\mathbb{M}}_{\text{sec}} \qquad (103)$$

Qualitatively, the interaction, Eq. (90), indicates that the larger the interaction tensor, the smaller the deviation of grain stresses with respect to the average stress. As a consequence, the tangent approximation tends to a uniform stress state or *lower-bound approximation* for $n \rightarrow \infty$ (Sachs, 1928). This rate-insensitive limit of the tangent formulation is an artefact created by the use of the above tangent-secant relation of the nonlinear polycrystal in the self-consistent solution of the linear comparison polycrystal. On the contrary, the secant interaction has been proven to tend to a uniform strain rate state or *upper-bound approximation* (Taylor, 1938) in the rate-insensitive limit.

4.5 Second-Order Formulation

The more accurate *second-order approximation* to linearize the behavior of the mechanical phase provides improved micromechanical predictions, which are obtained from the calculation of average fluctuations of the stress distribution inside the linearized SR grains. The methodology to obtain these fluctuations was derived by Bobeth and Diener (1987), Kreher (1990), and Parton and Buryachenko (1990) for composites and extended by Lebensohn et al. (2007) for polycrystals. The effective stress potential $\overline{U}_T$ of a linearly viscous polycrystal described by Eq. (65) may be written in the form (Laws, 1973; Willis, 1981),

$$\overline{U}_T = \frac{1}{2}\overline{\mathbb{M}} : (\boldsymbol{\Sigma}' \otimes \boldsymbol{\Sigma}') + \dot{\mathbf{E}}^0 : \boldsymbol{\Sigma}' + \frac{1}{2}\overline{G} \qquad (104)$$

where $\overline{G}$ is the power under zero applied stress. The self-consistent expression for $\overline{\mathbb{M}}$ and $\dot{\mathbf{E}}^0$, Eqs. (97) and (98), can be rewritten as

$$\overline{\mathbb{M}} = \langle \mathbb{M}^{(r)}\mathbb{B}^{(r)} \rangle = \sum_r c^{(r)}\mathbb{M}^{(r)}\mathbb{B}^{(r)} \qquad (105)$$

$$\dot{\mathbf{E}}^0 = \langle \mathbb{M}^{(r)}\mathbf{b}^{(r)} + \dot{\boldsymbol{\epsilon}}^{0(r)} \rangle = \sum_r c^{(r)} \left(\mathbb{M}^{(r)}\mathbf{b}^{(r)} + \dot{\boldsymbol{\epsilon}}^{0(r)} \right) = \sum_r c^{(r)} \dot{\boldsymbol{\epsilon}}^{0(r)} \mathbb{B}^{(r)} \tag{106}$$

where $c^{(r)}$ is the volume fraction associated with SR grain (r). The corresponding expression for $\overline{G}$ is

$$\overline{G} = \sum_r c^{(r)} \dot{\boldsymbol{\epsilon}}^{0(r)} : \mathbf{b}^{(r)}. \tag{107}$$

The average second-order moment of the stress field over a SR grain (r) of the polycrystal is a fourth-rank tensor given by:

$$\langle \boldsymbol{\sigma}' \otimes \boldsymbol{\sigma}' \rangle^{(r)} = \frac{2}{c^{(r)}} \frac{\partial \overline{U}_T}{\partial \mathbb{M}^{(r)}} \tag{108}$$

Replacing Eqs. (105)–(107) in (108), leads to

$$\langle \boldsymbol{\sigma}' \otimes \boldsymbol{\sigma}' \rangle^{(r)} = \frac{1}{c^{(r)}} \frac{\partial \overline{\mathbb{M}}}{\partial \mathbb{M}^{(r)}} : (\boldsymbol{\Sigma}' \otimes \boldsymbol{\Sigma}') + \frac{1}{c^{(r)}} \frac{\partial \dot{\mathbf{E}}^0}{\partial \mathbb{M}^{(r)}} : \boldsymbol{\Sigma}' + \frac{1}{c^{(r)}} \frac{\partial \overline{G}}{\partial \mathbb{M}^{(r)}} \tag{109}$$

The algorithmic expressions to calculate the partial derivatives on the right-hand side can be found in Lebensohn et al. (2007).

Once the average second-order moments of the stress field over each SR grain (r) are obtained, the implementation of the second-order procedure follows the work of Liu and Ponte Castañeda (2004). The covariance tensor of stress fluctuations in the SR grains of the linear comparison polycrystal is given by:

$$\mathbb{C}_{\sigma'}^{(r)} = \langle \boldsymbol{\sigma}' \otimes \boldsymbol{\sigma}' \rangle^{(r)} - \boldsymbol{\sigma}'^{(r)} \otimes \boldsymbol{\sigma}'^{(r)} \tag{110}$$

and the average and the average fluctuation of resolved shear stress on slip system k of SR grain (r) is given by

$$\overline{\tau}^{k(r)} = \mathbf{Z}_s^{k(r)} : \boldsymbol{\sigma}'^{(r)} \tag{111}$$

$$\hat{\tau}^{k(r)} = \overline{\tau}^{k(r)} \pm \left(\mathbf{Z}_s^{k(r)} : \mathbb{C}_{\sigma'}^{(r)} \mathbf{Z}_s^{k(r)} \right)^{1/2} \tag{112}$$

where the positive (negative) branch should be selected if $\overline{\tau}^{k(r)}$ is positive (negative). The slip potential associated with slip system k of the nonlinear polycrystal is defined as:

$$\phi^k(\tau) = \frac{\tau_0^k}{n+1} \left(\frac{|\tau|}{g^k} \right)^{n+1}. \tag{113}$$

The linearized local behavior associated with SR grain (r) is then given by:

$$\dot{\boldsymbol{\epsilon}}^{(r)} = \mathbb{M}_{SO}^{(r)}\boldsymbol{\sigma}'^{(r)} + \dot{\boldsymbol{\epsilon}}_{SO}^{0(r)} \tag{114}$$

with:

$$\mathbb{M}_{SO}^{(r)} = \sum_k \alpha^{k(r)}\left(\mathbf{Z}_s^{k(r)}\otimes\mathbf{Z}_s^{k(r)}\right) \tag{115}$$

$$\dot{\boldsymbol{\epsilon}}_{SO}^{0\,(r)} = \sum_k e^{k(r)}\mathbf{Z}_s^{k(r)} \tag{116}$$

where $\phi'^k(\tau) = d\phi^k/d\tau$ stand for two scalar magnitudes associated with each slip system k of each SR grain (r) are defined as

$$\alpha^{k(r)} = \frac{\phi'^{k(r)}\left(\hat{\tau}^{k(r)}\right) - \phi'^{k(r)}\left(\overline{\tau}^{k(r)}\right)}{\hat{\tau}^{k(r)} - \overline{\tau}^{k(r)}} \tag{117}$$

$$e^{k(r)} = \phi'^{k(r)}\left(\overline{\tau}^{k(r)}\right) - \alpha^{k(r)}\overline{\tau}^{k(r)} \tag{118}$$

Once the linear comparison polycrystal is defined by Eqs. (115) and (116), different second-order estimates of the effective behavior of the nonlinear aggregate can be obtained. Approximating the potential of the nonlinear polycrystal in terms of the potential of the linear comparison polycrystal and a suitable measure of the error, Liu and Ponte Castañeda (2004) generated the following expression (corresponding to the so-called *energy* version of the second-order theory) for the effective potential of the nonlinear polycrystal

$$\overline{U}(\boldsymbol{\Sigma}') = \sum_r c^{(r)} \sum_k \left\{\phi^{k(r)}\left(\hat{\tau}^{k(r)}\right) + \phi'^{k(r)}\left(\overline{\tau}^{k(r)}\right)\left(\overline{\tau}^{k(r)} - \hat{\tau}^{k(r)}\right)\right\} \tag{119}$$

from where the effective response of the homogenized polycrystal can be obtained as $\dot{\mathbf{E}} = \partial\overline{U}(\boldsymbol{\Sigma}')/\partial\boldsymbol{\Sigma}'$.

The alternative *constitutive equation* version of the second-order theory simply consists in making use of the effective stress–strain rate relations for the linear comparison polycrystal. The effective strain is obtained as

$$\dot{\mathbf{E}} = \sum_r c^{(r)} \sum_k \mathbf{Z}_s^{k(r)}\phi'^{k(r)}\left(\overline{\tau}^{k(r)}\right) \tag{120}$$

Both versions of the second-order theory give slightly different results, depending on nonlinearity and local anisotropic contrast. Such gap is relatively small compared with the larger differences obtained with the different

SC approaches. The *constitutive equation* version is—in principle—less rigorous since it does not derive from a potential function, but has the advantage that can be obtained by simply following the affine algorithm described in the previous sections, using the linearized moduli defined by Eqs. (115) and (116).

4.6 Numerical Implementation

To illustrate the practical use of the VPSC formulation, the steps required to predict the local and overall viscoplastic response of a polycrystal are detailed below. Starting for convenience with an initial Taylor guess, i.e., $\dot{\boldsymbol{\epsilon}}^{(r)} = \dot{\mathbf{E}}$ for all grains, the following nonlinear equation is solved to get $\boldsymbol{\sigma}'^{(r)}$,

$$\dot{\mathbf{E}} = \dot{\gamma}_0 \sum_k \mathbf{Z}_s^{k(r)} \left(\frac{\mathbf{Z}_s^{k(r)} : \boldsymbol{\sigma}'^{(r)}}{g^{k(r)}} \right)^n \operatorname{sign}\left(\mathbf{Z}_s^{k(r)} : \boldsymbol{\sigma}'^{(r)} \right) \tag{121}$$

and an appropriate first-order linearization scheme is used to obtain initial values of $\mathbb{M}^{(r)}$ and $\dot{\boldsymbol{\epsilon}}^{0(r)}$, for each SR grain ($r$). Next, initial guesses for the macroscopic moduli $\overline{\mathbb{M}}$ and $\dot{\mathbf{E}}^0$ are obtained (usually as simple averages of the local moduli). The initial guess for the macroscopic stress $\boldsymbol{\Sigma}'$ can be obtained from them and the applied strain rate, Eq. (65), while the Eshelby tensors $\mathbb{S}$ and $\mathbb{P}$ can be calculated using the macroscopic moduli and the ellipsoidal shape of the SR grains by means of the procedure described in Section 4.1. Subsequently, the interaction tensor $\tilde{\mathbb{M}}$, Eq. (91), and the localization tensors $\mathbb{B}^{(\mathrm{r})}$ and $\mathbf{b}^{(\mathrm{r})}$, Eqs. (94) and (95), can be calculated as well. With these tensors, new estimates of $\overline{\mathbb{M}}$ and $\dot{\mathbf{E}}^0$ are obtained by solving iteratively the self-consistent Eqs. (97) and (98). After achieving convergence on the macroscopic moduli (and, consequently, also on the macroscopic stress and the interaction and localization tensors), a new estimation of the average grain stresses can be obtained using the localization relation, Eq. (92). If the recalculated average grain stresses are different from the input values within certain tolerance, a new iteration should be started, until convergence is reached. If the chosen linearization scheme is the second-order formulation, an additional loop on the linearized moduli is needed, using the improved estimates of the second-order moments of the stress in the grains, obtained by means of the methodology described in Section 4.4. This additional loop roughly increases the calculation time by one order of magnitude with respect to first-order linearizations. When the iterative procedure is completed, the average shear rates on the slip system (*k*) in each grain (*r*) are calculated as:

$$\dot{\gamma}^{k(r)} = \dot{\gamma}_0 \left(\frac{\mathbf{Z}_s^{k(r)} : \boldsymbol{\sigma}'^{(r)}}{g^{k(r)}} \right)^n \operatorname{sign}\left(\mathbf{Z}_s^{k(r)} : \boldsymbol{\sigma}'^{(r)} \right) \tag{122}$$

These average shear rates are in turn used to calculate the lattice rotation rates associated with each SR grain:

$$\dot{\boldsymbol{\omega}}^{(r)} = \dot{\boldsymbol{\Omega}} + \dot{\tilde{\boldsymbol{\omega}}}^{(r)} - \dot{\boldsymbol{\omega}}^{p(r)} \tag{123}$$

where, following Eq. (61),

$$\dot{\boldsymbol{\omega}}^{p(r)} = \sum_k \mathbf{Z}_a^{k(r)} \dot{\gamma}^{k(r)} \tag{124}$$

where $\mathbf{Z}_a^{k(r)}$ is the (uniform) antisymmetric Schmid tensor of slip system *k* in SR grain (*r*).

The above numerical scheme can be used to predict the texture development by applying the viscoplastic deformation to the polycrystal in incremental steps. The latter is done by assuming constant rates during a time interval Δt (such that $\dot{\mathbf{E}}\Delta t$ corresponds to a macroscopic strain increment of the order of a few percent) and using: (a) the strain rates and rotation rates (times Δt) to update the shape and orientation of the SR grains, and (b) the shear rates (times Δt) to update the critical stress of the deformation systems due to strain hardening using one the models described in Section 3, after each deformation increment. Note that the above explicit update schemes relies on the fact that the orientation and hardening variables evolve slowly within the adopted time interval. Otherwise, Δt should be chosen smaller.

5. FULL-FIELD HOMOGENIZATION OF POLYCRYSTALS

Full-field homogenization aims to predict the macroscopic response and microscopic field distribution of heterogenous materials based on the numerical simulation of the mechanical response of an RVE of the material microstructure (Böhm, 2004). The method is computationally expensive because it involves the solution of a boundary value problem that might contain a large number of degrees of freedom. Nevertheless, it can provide more accurate predictions of the macroscopic response than mean-field homogenization models (such as VPSC) because it includes more accurate information of the details of the microstructure. In addition, full-field homogenization of polycrystals can account for the effect of microstructural details that cannot be easily included in mean-field models such as relative grain sizes and grain shape distribution, as well as grain boundary misorientation distributions. Finally, it provides very accurate information for the local values of the stress and strain fields as well as of the state variables throughout the microstructure. This information is critical for predicting damage localization and failure of heterogeneous materials.

Several methods have been used to simulate the mechanical response of an RVE in within the framework of full-field homogenization. Among them, the two most common approaches are the FFT-based method (Moulinec & Suquet, 1998), extended to viscoplastic polycrystals in Lebensohn (2001), and the FEM.

5.1 Homogenization Using the Finite Element Method

The first attempts to use the FEM to obtain the mechanical response of a polycrystal were carried out in the 1990s by Kalidindi, Bronkhorst, and Anand (1992), Bronkhorst, Kalidindi, and Anand (1992), and Miehe et al. (1999). These first works were limited to 2-D and the RVEs used for the simulations were just a regular arrangement of finite elements in which each element represented a whole grain. These studies were able to capture the macroscopic response and texture evolution under large strains but the distribution of local fields was missing due to the rudimentary representation of the microstructure. Indeed, these simulations were closer to mean-field polycrystalline models (such as Taylor or VPSC) than to full-field simulations because the fields in each grain were constant (or just vary linearly) and did not take into account the strong gradients found in actual polycrystals. The use of complex RVEs with more realistic representations of the polycrystalline microstructure started at the early 2000s (many years after the development of CP theory and of the first numerical implementations) due to the high computational cost of these type of models. The first 3-D simulations used regular arrangements of grains (for example, rhombic dodecahedra in Mika and Dawson (1999)) but each grain was represented using several finite elements and, thus, the strong deformation gradients due to the accommodation of the plastic strain incompatibility between adjacent grains were accounted for. Finally, the use of several elements per grain was combined with more complex representation of the microstructure in 3-D, leading to the first realistic full-field simulations (Barbe, Decker, Jeulin, & Cailletaud, 2001;

Barbe, Forest, & Cailletaud, 2001). Since then, full-field simulation using the FEM has been a very active area of research. The finite element implementation of CP models has been reviewed by Roters, Eisenlohr, Hantcherli, et al. (2010) and Roters, Eisenlohr, Bieler, and Raabe (2010) and therefore will not presented here. Only some notes on the boundary conditions used in the boundary value problem will be reviewed due to their implication in obtaining the macroscopic fields as result of the full-field simulations.

The boundary conditions used in the simulation of the RVE deformation are a key issue in full-field homogenization. These conditions are energetically consistent if the stress and strain microscopic fields and the corresponding macroscopic fields fulfill the so-called Hill-Mandel principle of macro-homogeneity (Hill, 1965b). This principle states that the macroscopic stress power should be equal to the volume average of the microscopic stress power. Four different types of boundary conditions can be imposed to the RVE resulting in microfields compatible with this condition (Böhm, 2004; Ostoja-Starzewski, 2006; Peric, de Souza Neto, Feijóo, Partovi, & Molina, 2011): (1) statically uniform boundary conditions where homogeneous surface tractions are applied to the RVE faces; (2) kinematically uniform boundary conditions where constant displacements are applied to the RVE boundary; (3) mixed conditions combining uniform tractions and constant displacements on the different RVE boundaries; and (4) periodic boundary conditions. For any given RVE, the results obtained under periodic boundary conditions (even in the case of nonperiodic microstructures) lie between the corresponding lower and upper estimates obtained with homogeneous boundary conditions (1) and (2) and show a faster convergence toward the actual effective response with increasing the RVE size (Hazanov & Huet, 1994). Moreover, periodic boundary conditions reproduce exactly the deformation of an infinite media constructed by a periodic arrangement of the RVE. For these reasons, periodic boundary conditions are the common choice for full-field homogenization and will be described here.

Periodic boundary conditions assume that the whole RVE deforms as a jigsaw puzzle. If a RVE is selected such that the periodicity of the microstructure is cubic, then the periodic translation of the cell in the three directions will fill the whole space (Fig. 4A). Let $\mathbf{l}_i = l_i\mathbf{e}_i$ be the three orthogonal vectors defining the cubic periodicity and let $\mathbf{e}_i$ be the corresponding unit vectors defining the basis. If x_1, x_2, x_3 are the coordinates of a point in the RVE in the system defined by $\mathbf{e}_i$, the periodic boundary conditions link

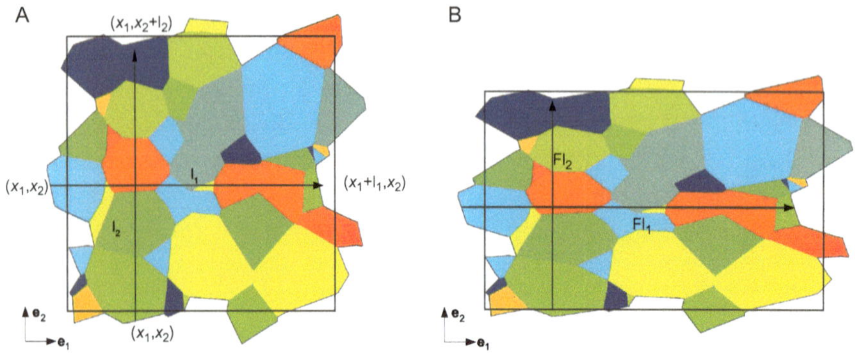

Fig. 4 Deformation of a periodic RVE of a polycrystal under periodic boundary conditions in 2-D. (A) Undeformed configuration and (B) Deformed shape under biaxial deformation.

the local displacement vector **u** of the nodes on opposite faces of the RVE with the far-field macroscopic deformation gradient $\overline{\mathbf{F}}$ according to,

$$\begin{aligned}\mathbf{u}(x_1, x_2, 0) - \mathbf{u}(x_1, x_2, L_3) &= (\overline{\mathbf{F}} - \mathbf{I})\mathbf{l}_3 \\ \mathbf{u}(x_1, 0, x_3) - \mathbf{u}(x_1, L_2, x_3) &= (\overline{\mathbf{F}} - \mathbf{I})\mathbf{l}_2 \\ \mathbf{u}(0, x_2, x_3) - \mathbf{u}(L_1, x_2, x_3) &= (\overline{\mathbf{F}} - \mathbf{I})\mathbf{l}_1.\end{aligned} \tag{125}$$

The resulting deformed cell preserves the cubic periodicity, but the new vectors defining the periodicity are given by $\overline{\mathbf{F}}\mathbf{l}_i$, as it can be observed in Fig. 4B.

These boundary conditions are implemented in the FEM by coupling the displacement degrees of freedom of the nodes in the RVE surface. The most direct implementation of periodic boundary conditions requires a periodic mesh, i.e., the 2-D discretization of two opposite faces of the RVE boundary should be identical, although some interpolation methods were proposed in the literature to overcome this requirement (Nguyen, Béchet, Geuzaine, & Noels, 2012). The application of Eq. (125) to the nodes of a periodic mesh can be done by constraint elimination (Michel, Moulinec, & Suquet, 1999; Peric et al., 2011) or by means of the Lagrange multiplier method (Böhm & Han, 2001; Segurado & LLorca, 2002). Standard finite element codes do not allow to apply direct constraint elimination and the common approach is then to use Lagrange multipliers by imposing a relation in the displacement of some nodes in the model (Segurado & LLorca, 2002, 2013). In particular, a master node M_i (i = 1, 2, 3) can be defined for each pair of opposite cube faces and the value of the far-field macroscopic deformation gradient is imposed to the RVE through the displacement of those master nodes according to

$$\mathbf{u}(M_i) = (\overline{\mathbf{F}} - \mathbf{I})\mathbf{l}_i. \tag{126}$$

If some components of the far-field deformation gradient are not known a priori (e.g., under uniaxial tension), the corresponding effective stresses $\overline{\boldsymbol{\sigma}}$ are set instead. This task is carried out by applying nodal forces P to the corresponding master node i and degree of freedom j according to

$$P_j(M_i) = (\overline{\boldsymbol{\sigma}}\mathbf{e}_i)_j A_i \tag{127}$$

where A_i is the area of the cell perpendicular to direction $\mathbf{e}_i$. Under small strains, the area A_i corresponds to the undeformed geometry and Eq. (127) is used to obtain the value of the forces to be applied as function of the target stress. If finite deformations are applied, A_i corresponds to the actual area and the relation between the force and the corresponding Cauchy stress component is not known a priori because the actual value of the area changes with the deformation. In this case, to impose a given history of any stress component, an iterative procedure should be applied to correct the applied forces with the cell deformation.

5.2 Homogenization Using the fast Fourier transform

The large number of degrees of freedom required by full-field homogenization limits the size of the microstructures that can be simulated using the FEM. Conceived as a very efficient alternative to the FEM, a numerical formulation based on the use of the FFT algorithm was originally proposed by Moulinec and Suquet (1994) to compute convolution integrals over the entire periodic RVE involving the periodic Green's function of a reference medium. In the original method, known as the basic scheme, the solution is obtained using a fixed-point iterative scheme. The contrast in phase properties has a strong impact in the convergence rate (the higher the contrast, the more numerically demanding the convergence). Moreover, the properties of the reference medium, although do not influence the final results, also strongly affect the convergence rate. Several modifications to the Moulinec–Suquet scheme have been proposed afterward (Eyre & Milton, 1999; Kabel, Böhlke, & Schneider, 2014; Michel et al., 2000; Monchiet & Bonnet, 2013; Zeman, Vondřejc, Novák, & Marek, 2010) based on modifying the iterative algorithm in order to improve the convergence rate. All these approaches rely on the use of a reference medium and, therefore, their performance is linked to an appropriate election of its properties.

An alternative approach to FFT-based homogenization in which the problem is derived using a weak formulation, analogous to finite elements, was recently proposed (de Geus, Vondrejc, Zeman, Peerlings, & Geers, 2017; Vondrejc, Zeman, & Marek, 2014; Zeman, de Geus, Vondrejc, Peerlings, & Geers, 2017). This approach does not require a reference medium and the Green's function of the classical schemes are replaced here by projection operators that ensure the compatibility of the strain fields. The similarities of this approach with Galerkin finite elements make very easy the introduction of generic nonlinear constitutive equations, including their corresponding tangents to accelerate the convergence.

FFT-based methods were originally developed by Moulinec and Suquet (1994) to carry out computational homogenization in periodic composites and have been profusely used for this class of materials (Brisard & Dormieux, 2010; Eyre & Milton, 1999; Idiart, Moulinec, & Ponte Castañeda, 2006; Michel et al., 1999; Moulinec & Silva, 2014; Moulinec & Suquet, 1998; Willot, 2015). In this type of problems, the source of heterogeneity is related to the spatial distribution of phases with different mechanical properties and the phase behavior is given in general by simple constitutive equations (in many cases, by isotropic linear elastic relations). The extension of FFT-based homogenization to polycrystals was proposed first in Lebensohn (2001) to exploit the superior numerical performance of CP-FFT with respect to CP-FE, allowing a large reduction of the computational time and/or the use of very complex and detailed polycrystalline RVEs. Since then, different numerical implementations of the FFT-based method for polycrystals have been developed, depending on the constitutive equation adopted to represent the behavior of each single crystal material point. Classic schemes have been used to solve the linearized problem. In almost all the cases, requiring the definition of a reference medium. An exception is the recent work by Lucarini and Segurado (2018) who adapted the Galerkin FFT-based approach to study the cyclic behavior and fatigue performance of polycrystals.

A summary of the type of problems and constitutive regimes in which FFT-based homogenization has been applied to polycrystals includes thermo-elasticity (Brenner, Lebensohn, & Castelnau, 2009), rigid-viscoplasticity (Lebensohn, 2001; Lebensohn et al., 2008, 2005, 2009), and elasto-viscoplasticity for small strains (Grennerat et al., 2012; Lebensohn, Kanjarla, & Eisenlohr, 2012) and large strains (Eisenlohr, Diehl, Lebensohn, & Roters, 2013; Lucarini & Segurado, 2018). More complex behaviors include dilatational plasticity (Lebensohn et al., 2013), strain gradient plasticity (Lebensohn & Needleman, 2016), curvature-driven plasticity (Upadhyay, Capolungo, Taupin, Fressengeas, & Lebensohn, 2016), and transformation plasticity (Otsuka, Brenner, & Bacroix, 2018; Richards, Lebensohn, & Bhattacharya, 2013). Recent implementations of field dislocation mechanics theories (Berbenni, Taupin, Djaka, & Fressengeas, 2014; Brenner, Beaudoin, Suquet, & Acharya, 2014) and discrete dislocation mechanics formulations (Bertin, Upadhyay, Pradalier, & Capolungo, 2015; Graham, Rollett, & Lesar, 2016) are also examples of the great potential of FFT-based formulations to provide the numerical efficiency needed for implementation of these very powerful, but also very complex and numerically demanding, emerging micromechanical formulations.

In what follows, the rigid-viscoplastic FFT-based formulation (VPFFT) for fully dense, incompressible polycrystals, and the dilatational extension of the latter (DVPFFT) to predict microstructural effects on porosity evolution in polycrystals with intergranular voids are presented in order to illustrate the methodology and provide the theory and algorithms behind one of the examples of Section 7. For details of other numerical implementations of the FFT-based formulation, the reader is referred to the specific papers listed above.

Briefly, the viscoplastic FFT-based formulation consists in iteratively adjusting a compatible strain rate field, related with an equilibrated stress field through a constitutive potential, such that the average of local work rates is minimized. The method is based on the fact that the local mechanical response of a heterogeneous medium can be calculated as a convolution integral between Green functions associated with appropriate fields of a linear reference homogeneous medium and the actual heterogeneity field. For a periodic medium, use can be made of the discrete Fourier transform to reduce convolution integrals in real space to simple products in Fourier space. Thus, the FFT algorithm can be utilized to transform the heterogeneity field into Fourier space and, in turn, to get the mechanical fields by transforming that product back to real space. However, the actual heterogeneity field depends precisely on the a priori unknown mechanical fields. Therefore, an iterative scheme has to be implemented to obtain, upon convergence, a compatible strain rate field and a stress field in equilibrium.

A periodic RVE of the polycrystal is discretized into $N_1 \times N_2 \times N_3$ Fourier points. This discretization determines a regular grid in the Cartesian space $\{\mathbf{x}^d\}$ and a corresponding grid in Fourier space $\{\boldsymbol{\xi}^d\}$. Velocities and tractions along the boundary of the unit cell are left undetermined.

A velocity gradient $V_{i,j}$ (which can be decomposed into a symmetric strain rate and a antisymmetric rotation rate: $V_{i,j} = \dot{E}_{ij} + \dot{\Omega}_{ij}$) is imposed to the unit cell. The local strain rate field is a function of the local velocity field, i.e., $\dot{\epsilon}_{ij}(v_k(\mathbf{x}))$, and can be split into its average and a fluctuation term: $\dot{\boldsymbol{\epsilon}}_{ij}(v_k(\mathbf{x})) = \dot{E}_{ij} + \tilde{\dot{\epsilon}}_{ij}(\tilde{v}_k(\mathbf{x}))$, where $v_i(\mathbf{x}) = \dot{E}_{ij}x_j + \tilde{v}_i(\mathbf{x})$. By imposing periodic boundary conditions, the velocity fluctuation field $\tilde{v}_k(\mathbf{x})$ is assumed to be periodic across the boundary of the unit cell, while the traction field is antiperiodic, to meet equilibrium on the boundary between contiguous unit cells.

The local constitutive relation between the strain rate $\dot{\boldsymbol{\epsilon}}(\mathbf{x})$ and the deviatoric stress $\boldsymbol{\sigma}'(\mathbf{x})$ is given by the same rate-sensitivity relation used within the VPSC framework, Eq. (59). If a fourth-rank tensor $\mathbb{L}^0$ is chosen as the stiffness of a linear reference medium (the choice of $\mathbb{L}^0$ can be quite arbitrary, but the speed of convergence of the method will depend on this choice). The polarization field $\varphi_{ij}(\mathbf{x})$, Eq. (70), is defined as

$$\varphi_{ij}(\mathbf{x}) = \tilde{\sigma}'_{ij}(\mathbf{x}) - L^0_{ijkl}\tilde{\dot{\epsilon}}_{kl}(\mathbf{x}) \tag{128}$$

Then, the Cauchy stress deviation can be written as

$$\tilde{\sigma}_{ij}(\mathbf{x}) = L^0_{ijkl}\tilde{\dot{\epsilon}}_{kl}(\mathbf{x}) + \varphi_{ij}(\mathbf{x}) + \tilde{\sigma}^m(\mathbf{x})\delta_{ij} \tag{129}$$

and combining Eq. (129) with the equilibrium ($\sigma_{ij,j}(\mathbf{x}) = \tilde{\sigma}_{ij,j}(\mathbf{x}) = 0$), the incompressibility condition, and the relation $\tilde{\dot{\epsilon}}_{ij}(\mathbf{x}) = \frac{1}{2}\left(\tilde{v}_{i,j}(\mathbf{x}) + \tilde{v}_{j,i}(\mathbf{x})\right)$ leads to,

$$\left| \begin{array}{l} L^0_{ijkl}\tilde{v}_{k,lj}(\mathbf{x}) + \tilde{\sigma}^m_{,i}(\mathbf{x}) + \varphi_{ij,j}(\mathbf{x}) = 0 \\ \tilde{v}_{k,k}(\mathbf{x}) = 0 \end{array} \right. \tag{130}$$

This system of differential equations is formally equivalent to system (69). However, both systems actually differ in that: (a) the HEM's stiffness modulus $\overline{\mathbb{L}}$ of Eq. (69) is replaced in Eq. (130) by the stiffness of a linear reference medium $\mathbb{L}^0$, and (b) the polarization field in Eq. (130) has in general non-vanishing values throughout the unit cell and is periodic (owing to the RVE periodicity), while the polarization field in Eq. (69) vanishes outside the domain of the inclusion. The auxiliary system involving periodic Green functions is, then, given by Eq. (73),

$$\left| \begin{array}{l} L^0_{ijkl}\, G_{km,lj}(\mathbf{x}-\mathbf{x}') + H_{m,i}(\mathbf{x}-\mathbf{x}') + \delta_{im}\delta(\mathbf{x}-\mathbf{x}') = 0 \\ G_{km,k}(\mathbf{x}-\mathbf{x}') = 0 \end{array} \right. \tag{131}$$

After some manipulation, the velocity and the velocity gradient deviation fields are given by the convolution integrals

$$\tilde{v}_k(\mathbf{x}) = \int_{\mathbb{R}^3} \mathrm{G_{ki,j}}(\mathbf{x}-\mathbf{x}')\ \varphi_{\mathrm{ij}}(\mathbf{x}')\ d\mathbf{x}' \tag{132}$$

$$\tilde{v}_{i,j}(\mathbf{x}) = \int_{\mathbb{R}^3} \mathrm{G_{ik,jl}}(\mathbf{x}-\mathbf{x}')\ \varphi_{\mathrm{kl}}(\mathbf{x}')\ d\mathbf{x}'. \tag{133}$$

Convolution integrals in direct space are simply products in Fourier space

$$\hat{\tilde{v}}_k(\boldsymbol{\xi}) = (-i\xi_j)\ \hat{G}_{ki}(\boldsymbol{\xi})\ \hat{\varphi}_{ij}(\boldsymbol{\xi}) \tag{134}$$

$$\hat{\tilde{v}}_{i,j}(\boldsymbol{\xi}) = \hat{\Gamma}_{ijkl}(\boldsymbol{\xi})\ \hat{\varphi}_{kl}(\boldsymbol{\xi}) \tag{135}$$

where the symbol "∧" indicated a Fourier transform. The Green operator in Eq. (135) is defined as $\Gamma_{ijkl} = G_{ik,jl}$.

The tensors $\hat{G}_{ij}(\boldsymbol{\xi})$ and $\hat{\Gamma}_{ijkl}(\boldsymbol{\xi})$ can be calculated by taking Fourier transform to the set of Eq. (131)

$$\left| \begin{array}{l} \xi_l\xi_j\ L^0_{ijkl}\hat{G}_{km}(\boldsymbol{\xi}) + i\xi_i\hat{H}_m(\boldsymbol{\xi}) = \delta_{im} \\ \xi_k\hat{G}_{km}(\boldsymbol{\xi}) = 0 \end{array} \right. \tag{136}$$

Defining the 3 × 3 matrix $A'_{ik}(\xi) = \xi_l\xi jL^0_{ijkl}$ and embedding it in the 4 × 4 matrix $A(\xi)$ as

$$A(\xi) = \begin{vmatrix} A'_{11} & A'_{12} & A'_{13} & \xi_1 \\ A'_{21} & A'_{22} & A'_{23} & \xi_2 \\ A'_{31} & A'_{32} & A'_{33} & \xi_3 \\ \xi_1 & \xi_2 & \xi_3 & 0 \end{vmatrix} \tag{137}$$

it leads to (see Eqs. 74–76)

$$\hat{G}_{ij}(\boldsymbol{\xi}) = A^{-1}_{ij}\ (i,j=1,3) \tag{138}$$

$$\hat{\Gamma}_{ijkl}(\boldsymbol{\xi}) = -\xi_j\xi_l\hat{G}_{ik}(\boldsymbol{\xi}) \tag{139}$$

5.3 Numerical Implementation

Assigning an initial guess values to the strain rate field in the regular grid $\{\mathbf{x}^d\}$ (e.g., $\tilde{\dot{\epsilon}}^0_{ij}(\mathbf{x}^d) = 0 \Rightarrow \dot{\epsilon}^0_{ij}(\mathbf{x}^d) = \dot{E}_{ij}$), and computing the corresponding stress field $\sigma'^0_{ij}(\mathbf{x}^d)$ from the local constitutive relation, Eq. (59) (which requires to know the initial values of the critical stresses $g^k(\mathbf{x}^d)$, and the Schmid tensors $\mathbf{Z}_a(\mathbf{x}^d)$), an initial guess for the polarization field in direct space $\varphi^0_{ij}(\mathbf{x}^d)$, Eq. (59), can be obtained which in turn can be Fourier-transformed to obtain $\hat{\varphi}^0_{ij}(\boldsymbol{\xi}^d)$. Furthermore, assuming that $\lambda^0_{ij}(\mathbf{x}^d)=\sigma^0_{ij}(\mathbf{x}^d)$ is the initial guess for a field of Lagrange multipliers associated with the compatibility constraints, the iterative procedure based on Augmented Lagrangians proposed by Michel et al. (2000) can be used. If the polarization field after iteration n is known, the $n+1$-th iteration starts by computing the new guess for the kinematically admissible strain rate deviation field

$$\hat{\tilde{d}}^{n+1}_{ij}(\boldsymbol{\xi}^d)=-\hat{\Gamma}^{sym}_{ijkl}(\boldsymbol{\xi}^d)\,\hat{\varphi}^n_{kl}(\boldsymbol{\xi}^d),\ \ \forall\boldsymbol{\xi}^d\neq\mathbf{0};\ \text{and}\ \ \hat{\tilde{d}}^{n+1}_{ij}(\mathbf{0})=\mathbf{0} \tag{140}$$

where $\hat{\Gamma}^{sym}_{ijkl}$ is the Green operator, appropriately symmetrized. The corresponding field in real space is, thus, obtained by application of the inverse FFT, i.e.,

$$\tilde{d}^{n+1}_{ij}(\mathbf{x}^d)=\mathrm{fft}^{-1}\left\{\hat{\tilde{d}}^{n+1}_{ij}(\boldsymbol{\xi}^{\mathbf{d}})\right\} \tag{141}$$

and the new guess for the deviatoric stress field is calculated from (omitting subindices):

$$\begin{aligned}\boldsymbol{\sigma}'^{n+1}(\mathbf{x}^d)+\mathbb{L}^0:\dot{\gamma}_0\sum_k\mathbf{Z}^k_s(\mathbf{x}^d)\left(\frac{\mathbf{Z}^k_s(\mathbf{x}^d):\boldsymbol{\sigma}'^{n+1}(\mathbf{x}^d)}{g^k(\mathbf{x}^d)}\right)^n\mathrm{sign}\left(\mathbf{Z}^k_s(\mathbf{x}^d):\boldsymbol{\sigma}'^{n+1}(\mathbf{x}^d)\right)\\=\boldsymbol{\lambda}^n(\mathbf{x}^d)+\mathbb{L}^0:(\dot{\mathbf{E}}+\tilde{\boldsymbol{d}}^{n+1}(\mathbf{x}^d))\end{aligned} \tag{142}$$

The iteration is completed with the calculation of the new guess for the Lagrange multiplier field according to

$$\boldsymbol{\lambda}^{n+1}(\mathbf{x}^d)=\boldsymbol{\lambda}^n(\mathbf{x}^d)+\mathbb{L}^0:(\tilde{\dot{\boldsymbol{\epsilon}}}^{n+1}(\mathbf{x}^d)-\tilde{\boldsymbol{d}}^{n+1}(\mathbf{x}^d)) \tag{143}$$

Eqs. (142) and (143) guarantee the convergence of $\dot{\boldsymbol{\epsilon}}(\mathbf{x}^d)$ (i.e., the strain rate field related with the stress through the constitutive equation) toward $\boldsymbol{d}(\mathbf{x}^d)$ (i.e., the kinematically admissible strain rate field) to fulfill compatibility, and of the Lagrange multiplier field $\boldsymbol{\lambda}(\mathbf{x}^d)$ toward the stress field $\boldsymbol{\sigma}'(\mathbf{x}^d)$ to fulfill equilibrium.

5.4 Time Integration

Upon convergence, the microstructure can be updated using an explicit scheme. The resulting shear strain rate field

$$\dot{\gamma}^k(\mathbf{x}^d) = \dot{\gamma}_0 \left(\frac{\mathbf{Z}_s^k(\mathbf{x}^d) : \boldsymbol{\sigma}'(\mathbf{x}^d)}{g^k(\mathbf{x}^d)} \right)^n \text{sign}\left(\mathbf{Z}_s^k(\mathbf{x}^d) : \boldsymbol{\sigma}'(\mathbf{x}^d)\right) \tag{144}$$

can be assumed to be constant during a time interval $[t, t + \Delta t]$. The macroscopic and local strain increments are then calculated as $\Delta E_{ij} = \dot{E}_{ij}\Delta t$ and $\Delta\epsilon_{ij}(\mathbf{x}^d) = \dot{\epsilon}_{ij}(\mathbf{x}^d)\Delta t$ and the local crystallographic orientations are updated according to the following local lattice rotation

$$\omega_{ij}(\mathbf{x}^d) = \left(\dot{\Omega}_{ij}(\mathbf{x}^d) + \dot{\tilde{\omega}}_{ij}(\mathbf{x}^d) - \dot{\omega}_{ij}^p(\mathbf{x}^d)\right)\Delta t \tag{145}$$

where $\dot{\omega}_{ij}^p(\mathbf{x}^d)$ can be obtained from Eqs. (61) and (176), and $\dot{\tilde{\boldsymbol{\omega}}}(x^d)$ is obtained back-transforming the converged antisymmetric field

$$\hat{\dot{\tilde{\omega}}}_{ij}(\boldsymbol{\xi}^d) = -\hat{\Gamma}_{ijkl}^{antisym}(\boldsymbol{\xi}^d)\,\hat{\varphi}_{kl}(\boldsymbol{\xi}^d), \quad \forall \boldsymbol{\xi}^d \neq 0; \text{ and } \hat{\dot{\tilde{\omega}}}_{ij}(\mathbf{0}) = \mathbf{0}. \tag{146}$$

The CRSSs of the deformation systems associated with each material point should also be updated after each deformation increment due to strain hardening using one the models described in Section 3.

After each time increment, the new position of the Fourier points can be determined calculating the velocity fluctuation term $\tilde{v}_k(\mathbf{x}^d)$ back-transforming Eq. (132), and

$$X_i(\mathbf{x}^d) = x_i^d + \left(\dot{E}_{ij}x_j^d + \tilde{v}_i(\mathbf{x}^d)\right)\Delta t. \tag{147}$$

Evidently, the set of advected Fourier points no longer forms a regular grid, after the very first deformation increment due to the heterogeneity of the medium. A rigorous way of dealing with this situation was proposed by Lahellec, Michel, Moulinec, and Suquet (2001) based on the particle-in-cell method (Sulsky, Zhou, & Schreyer, 1995).

5.5 Extension to Dilatational Viscoplasticity

The previous FFT-based algorithm can be adapted to the case of dilatational viscoplasticity of voided materials with polycrystalline or homogeneous isotropic matrix. The constitutive behavior of Fourier points belonging to a void is simply $\boldsymbol{\sigma} = \mathbf{0}$. In the cases of porous materials with crystalline or isotropic matrix, the constitutive equation for Fourier points with material properties is given, respectively, by Eq. (59), or by

$$\dot{\boldsymbol{\epsilon}}(\boldsymbol{x}^d) = \frac{3\dot{\gamma}_0}{2\sigma_0}\left(\frac{\sigma_{eq}(\mathbf{x}^d)}{\sigma_0}\right)^{n-1}\boldsymbol{\sigma}(\mathbf{x}^d) \tag{148}$$

where σ_0 is the flow stress.

Given the effective compressibility of the material due to the presence of voids, the differential equation whose solution is the Green's function $G_{km}(\mathbf{x}^d)$ associated with the velocity field is given by Eq. (131)

$$L^o_{ijkl} G_{km,lj}(\mathbf{x}-\mathbf{x}') + \delta_{im}\,\delta(\mathbf{x}-\mathbf{x}') = \mathbf{0} \tag{149}$$

hence (see Eqs. 138 and 139)

$$\hat{G}_{ij}(\boldsymbol{\xi}) = A^{-1}_{ij}(\boldsymbol{\xi}) \text{ and } A_{ik}(\boldsymbol{\xi}) = \xi_l \xi_j L^o_{ijkl} \tag{150}$$

and

$$\hat{\Gamma}_{ijkl}(\boldsymbol{\xi}) = -\xi_j \xi_l \hat{G}_{ik}(\boldsymbol{\xi}). \tag{151}$$

The algorithm for a full stress tensor $\boldsymbol{\Sigma}$ imposed to a unit cell representing a porous material requires an initial guess for the average strain rate

$$\dot{E}^0_{ij} = \dot{E}'^0_{ij} + \frac{\dot{E}^0_{kk}}{3}\delta_{ij} \tag{152}$$

where $\dot{E}'_{ij}$ and $\dot{E}_{kk}$ are the deviatoric and hydrostatic parts, which will be adjusted iteratively. Initial guess values also need to be assigned to the local strain rate field. The deviatoric part is taken to be uniform

$$\tilde{\dot{\epsilon}}'^0_{ij}(\mathbf{x}) = \mathbf{0} \Rightarrow \dot{\epsilon}'^0_{ij}(\mathbf{x}) = \dot{E}'^0_{ij}. \tag{153}$$

An initial guess of uniform dilatation is assumed for the Fourier points representing voids,

$$\tilde{\dot{\epsilon}}^o_{kk}(\mathbf{x}) = \left(\frac{1}{f} - 1\right)\dot{E}^o_{kk} \tag{154}$$

where f is the current porosity in the RVE. For these initial values, the corresponding stress field in the material points $\boldsymbol{\sigma}^0(\mathbf{x})$ is obtained by inverting the local constitutive relation given by Eq. (59) (for polycrystals) or Eq. (148) (for isotropic matrix), whereas for points belonging to voids, the stress simply vanishes. The initial specification of these fields and reference stiffness allows us to calculate the initial guess for the polarization field in direct space, Eq. (128), which in turn is Fourier-transformed, used in combination with Eq. (151) to solve the convolution integral in Fourier space, and antitransformed to obtain a new guess for the strain rate field, etc. Furthermore, assuming $\lambda^o_{ij}(\mathbf{x}) = \sigma^o_{ij}(\mathbf{x})$ as the initial guess for an auxiliary stress field associated with the compatibility constraint, the Augmented Lagrangian procedure (Eqs. 142 and 143) can be readily applied to iteratively determine the local fields. Microstructural changes, including porosity evolution, can then be tracked using the updating algorithms described in Lebensohn et al. (2013).

6. MULTISCALE MODELING OF POLYCRYSTALS

Simulation of polycrystals by means of either mean-field homogenization schemes or full-field homogenization of an RVE of the microstructure can only consider homogeneous stress or strain boundary conditions. Thus, it cannot be used to determine the mechanical response of polycrystals with arbitrary geometry or subjected to complex mechanical loads. Multiscale frameworks in which the single crystal (microscopic) behavior is connected with the specimen (macroscopic) response through an intermediate (mesoscopic) scale are the most direct solution to link the nonhomogeneous deformation of a structure with its microstructure. The behavior at mesoscopic scale is obtained from the homogenized response of a polycrystal that represents the behavior of each (polycrystalline) material point in the macroscale. These models allow considering microstructural effects on the response of a polycrystalline specimen and they are only applicable to problems with a clear separation of scales, in which the length-scale associated with the gradients of the mechanical fields at the macroscale has to be large compared with the length-scale of the polycrystalline microstructure, i.e., the grain or subgrain size.

The models and numerical methods at both ends of this multiscale problem are well-established and readily available. The mechanical behavior of single crystals is modeled using any of the CP models presented in Section 3, while the macroscopic behavior at the macroscopic scale, possibly of complicated geometry and undergoing complex boundary conditions, is modeled with the FEM using large-strain kinematics. Different flavors of the multiscale simulation strategy are determined by the model used at the mesoscopic scale. For example, they can be based on full-field homogenization, in which the mechanical response at each integration point is given by numerical simulation of a polycrystalline RVE associated to that integration point. Within this group, finite elements have been used to perform this mesoscale computational homogenization for different polycrystalline microstructures (Kouznetsova & Geers, 2008; Miehe et al., 1999; Werwer & Cornec, 2000). In the general case of heterogeneous materials (of which a polycrystal is a particular case), these approaches are sometimes referred to as FE^2 models (Feyel, 2003).

However, FEM-based homogenization at the mesoscale requires a huge computational effort for the whole multiscale calculation. An obvious direction for reducing this cost is the use of more efficient full-field models based on FFT to obtain the average behavior of the RVE in each integration point (Section 5.2). A more complex alternative is the use of model order reduction for the mecanical behavior at the microscale. Model order reduction has been mainly used for linear heterogeneous materials or more simple constitutive models at the microscale, but it has been recently applied to polycrystals (Michel & Suquet, 2016). Finally, the computational cost of concurrent multiscale models is strongly reduced if, instead of a numerical model, a mean-field homogenization approach (Section 4) is used at the microscale. In this latter case, the polycrystal response at each integration point is obtained from an estimate of its effective behavior, based a nonspace-resolved statistical representation of the microstructure and homogenization theory to connect the single crystal and polycrystal levels.

Another important aspect of the implementation at mesoscopic level within the multiscale analysis, as discussed by Van Houtte, Kanjarla, Van Bael, Seefeldt, and Delannay (2006), concerns whether the polycrystal model is embedded in the FE computation (i.e., interrogated on the fly each time the behavior of the corresponding material point is needed by the FE analysis) or used in a hierarchical fashion (i.e., the FE model utilizes a number of parameters which are identified in advance (precomputed), using the mesoscale model) (Van Houtte et al., 2006; Van Houtte, Yerra, & Van Bael, 2009). Embedded models are obviously more accurate to follow the microstructural evolution and how the later affects the macroscopic behavior, while hierarchical approximations are more efficient. In connection with these two different strategies, it is also worth mentioning a promising intermediate approach, known as the adaptive sampling method (Barton et al., 2008), which consists in embedding a lower length-scale model (e.g., at polycrystal level) and storing the response obtained for a given microstructure each time this model is interrogated. In this way, a microstructure-response database is populated, which can be used instead of a new call to the polycrystal model when the current microstructure is not far from an already computed point of this microstructure-response space.

Currently, one of the most efficient multiscale approaches is based in the combination of FEM at the macroscopic level with a VPSC approximation at the mesoscale. Along these lines, Segurado, Lebensohn, Llorca, and Tomé (2012) developed a semiimplicit elasto-visco-plastic multiscale framework to simulate the plastic deformation of polycrystalline structures.

This modeling framework have been later extended for different type crystals and applied to simulate thermomechanical processes on full components (Knezevic & Beyerlein, 2018; Knezevic, Crapps, et al., 2016; Zecevic, Beyerlein, & Knezevic, 2017). Due to the relevance of this work, the main equations of the framework developed in Segurado et al. (2012) will be reviewed here, including some implementation aspects in a standard finite element code.

6.1 Kinematics

The mesoscopic material model based on VPSC (Segurado et al., 2012) was developed using both small-strain and finite deformation frameworks. For finite deformations, a *hypoelastic* constitutive formulation was used. Hypoelastic models are a common approach to extend small-strain plasticity models to the finite strain range by recasting the original evolution equations in terms of objective stress rates that preserve frame invariance (de Souza Neto, Peric, & Owen, 2008). This approach to define user materials is a common option in most of commercial codes as ABAQUS, i.e., the one chosen for the numerical implementation of the model, due to its simplicity and usability. The application of the hypoelastic framework in this multiscale model allowed using small-strain kinematics inside the VPSC-based user-defined material subroutine in combination with a corotational reference frame that is determined and stored to account for large strains and rotations. In small strains, the total strain increment ($\Delta\boldsymbol{\epsilon}$) is decomposed in elastic ($\Delta\boldsymbol{\epsilon}_{el}$) and viscoplastic ($\Delta\boldsymbol{\epsilon}_{vp}$) parts, such that the constitutive relation at each polycrystalline integration point is given by:

$$\Delta\boldsymbol{\epsilon} = \Delta\boldsymbol{\epsilon}_{el} + \Delta\boldsymbol{\epsilon}_{vp} = \mathbb{L}^{-1}\Delta\boldsymbol{\sigma} + \Delta\boldsymbol{\epsilon}_{vp} \tag{155}$$

where $\mathbb{L}$ is the elastic stiffness tensor of the polycrystalline material point, $\Delta\boldsymbol{\sigma}$ is the stress increment and $\Delta\boldsymbol{\epsilon}_{vp} = \Delta\boldsymbol{\epsilon}_{vp}(\boldsymbol{\sigma})$ is computed with the VPSC model for each polycrystalline material point. For large deformations, the stress increment is based on the interval integration of the Jaumann rate of the Cauchy stress. In consistency with ABAQUS's treatment of rotations.

A *local coordinate system*, defined by the unit vectors $\mathbf{e}_\alpha$ $(\alpha = 1, 3)$ in which the polycrystal behavior is computed, undergoes rotations that need to be accounted for by the FEM. This system will be referred to as *corotational system*. In the adopted hypoelastic solution of the finite deformation problem, the material model is formulated incrementally, using the Jaumann rate of the Cauchy stress as the objective rate to define the stress increments. In this case, the evolution of the unit vectors of the local coordinate system rotating with the polycrystal is given by:

$$\dot{\mathbf{e}}_{\alpha} = \dot{\boldsymbol{\omega}}\,\mathbf{e}_{\alpha} \tag{156}$$

where $\dot{\boldsymbol{\omega}}$ is the total rotation rate of the polycrystalline material point. The integration in a time increment (from time t to $t + \Delta t$) gives an incremental rotation matrix $\Delta\mathbf{R}$ that rotates the local reference system from t to $t + \Delta t$,

$$\mathbf{R}^{t+\Delta t} = \Delta\mathbf{R}\mathbf{R}^{t} \tag{157}$$

where $\mathbf{R}^{t+\Delta t}$ is the rotation matrix from the local coordinate system at $t = 0$ to its orientation at time $t + \Delta t$, e.g.,

$$\mathbf{e}_{\alpha}^{t+\Delta t} = \mathbf{R}^{t+\Delta t}\mathbf{e}_{\alpha}^{0} \tag{158}$$

Similarly to Eq. (155), the velocity gradient $\mathbf{L}$ for large strains is decomposed into elastic and viscoplastic components

$$\mathbf{L} = \mathbf{L}_{el} + \mathbf{L}_{vp} \tag{159}$$

where

$$\mathbf{L}_{el} = \dot{\boldsymbol{\epsilon}}_{el} + \dot{\boldsymbol{\omega}} \tag{160}$$

and

$$\mathbf{L}_{vp} = \dot{\boldsymbol{\epsilon}}_{vp} \tag{161}$$

Elastic and viscoplastic constitutive laws involve only the symmetric component of the deformation, while the antisymmetric component is ambiguous. Consequently, the full rotation rate in Eq. (160) was assigned to the elastic part of the velocity gradient for convenience.

The Jaumann rate of the Kirchhoff stress, $\overset{\nabla}{\boldsymbol{\tau}}$, which is the stress measure work-conjugate with the rate of deformation, is defined by

$$\overset{\nabla}{\boldsymbol{\tau}} = \dot{\boldsymbol{\sigma}} - \dot{\boldsymbol{\omega}}\boldsymbol{\sigma} + \boldsymbol{\sigma}\dot{\boldsymbol{\omega}}. \tag{162}$$

If plastic strains are much larger than elastic ones, deformation can be considered isochoric ($J = \det(\mathbf{F}) = 1$ and $\boldsymbol{\tau} = J\boldsymbol{\sigma} \approx \boldsymbol{\sigma}$). Thus, $\overset{\nabla}{\boldsymbol{\sigma}}$ is related to the elastic strain rate by

$$\overset{\nabla}{\boldsymbol{\sigma}} = \mathbb{L}\left(\dot{\boldsymbol{\epsilon}} - \dot{\boldsymbol{\epsilon}}_{vp}\right). \tag{163}$$

The FE solution requires that the constitutive relation in Eq. (163) is expressed and integrated in a fixed coordinate system. If the coordinate system associated to every polycrystalline material point is unique and given by $\mathbf{e}_\alpha^0$, this reference frame is used as the required fixed coordinate system. Integrating Eq. (163) from t to $t + \Delta t$ gives

$$\overset{\nabla}{\boldsymbol{\sigma}}\Delta t = \mathbb{L}\left(\Delta\boldsymbol{\epsilon} - \Delta\boldsymbol{\epsilon}_{vp}\right) \tag{164}$$

In this formulation, the polycrystal elastic stiffness and the viscoplastic strain increment are calculated by means of the VPSC model in the corotational system. Therefore, use is made of $\mathbf{R}^{t+\Delta t}$ to express these magnitudes in the fixed coordinate system, e.g., if $\Delta\boldsymbol{\epsilon}_{vp}^*$ is the viscoplastic strain increment computed by means of the VPSC model (the symbol "*" denotes magnitudes expressed in the corotational system), $\Delta\boldsymbol{\epsilon}_{vp}$ in Eq. (164) is obtained as

$$\Delta\boldsymbol{\epsilon}_{vp} = \left(\mathbf{R}^{t+\Delta t}\right)^T \Delta\boldsymbol{\epsilon}_{vp}^* \mathbf{R}^{t+\Delta t}. \tag{165}$$

In Eq. (164), $\Delta\boldsymbol{\epsilon}$ is the logarithmic strain increment, obtained from $\Delta\boldsymbol{\epsilon}^*$ given by the FE analysis as

$$\Delta\boldsymbol{\epsilon}^* = \log(\Delta\mathbf{V}) \tag{166}$$

with $\Delta\mathbf{V}$ being the left Cauchy stretch tensor associated with the deformation gradient increment from t to $t + \Delta t$, such that $\Delta\mathbf{F} = \Delta\mathbf{V}\Delta\mathbf{R}$. Hence,

$$\Delta\boldsymbol{\epsilon} = \left(\mathbf{R}^{t+\Delta t}\right)^T \Delta\boldsymbol{\epsilon}^* \mathbf{R}^{t+\Delta t}. \tag{167}$$

Finally, for completeness, once the right hand of Eq. (164) is obtained by computing the left-hand magnitudes in the fixed reference system, the stress expressed in corotational axes (as needed by VPSC as part of the proposed algorithm, see next section) is obtained as

$$\boldsymbol{\sigma}^{t+\Delta t*} = \mathbf{R}^{t+\Delta t}\left(\boldsymbol{\sigma}^t + \overset{\nabla}{\boldsymbol{\sigma}}\Delta t\right)\left(\mathbf{R}^{t+\Delta t}\right)^T. \tag{168}$$

6.2 VPSC-Based UMAT Implementation

The approach at the macroscopic level in the FEM implementation is standard: the applied load is divided in increments, and the equilibrium at each increment is obtained by means of the FEM analysis in an iterative fashion, using a global nonlinear solver. The load increment is controlled by the time, and once the problem has being solved at time t, the solution for the next time increment requires the polycrystal model to provide a tangent stiffness

(Jacobian) matrix $\mathbb{L}^{tg} = \partial\Delta\boldsymbol{\sigma}/\partial\Delta\boldsymbol{\epsilon}$ for each material point, so the FEM can compute an initial guess for the nodal displacements at $t + \Delta t$. The strain increments obtained from that prediction for each material point, $\Delta\boldsymbol{\epsilon}^{FE}$, together with the stress $\boldsymbol{\sigma}^t$ and the set of internal state variables q_i^t corresponding to the previous increment, are used inside the UMAT to calculate a new guess for the stress and the Jacobian at $t + \Delta t$. When convergence in stress equilibrium is achieved by the global nonlinear scheme, the new values (at $t + \Delta t$) of the stresses, the internal variables, and the Jacobian matrix are accepted for every node, and the calculation advances to the next increment. For a given $\Delta\boldsymbol{\epsilon}^{FE}$, the VPSC-UMAT is based on the minimization of an ad hoc residual, defined below.

$$\boldsymbol{\sigma}^{t+\Delta t} = \boldsymbol{\sigma}^t + \mathbb{L}\Delta\boldsymbol{\epsilon}_{el} = \boldsymbol{\sigma}^t + \mathbb{L}\left(\Delta\boldsymbol{\epsilon} - \Delta\boldsymbol{\epsilon}_{vp}\right) \tag{169}$$

where $\mathbb{L}$ is the elastic stiffness of the polycrystal. In the present context, the natural choice for $\mathbb{L}$ is to use the elastic self-consistent (ELSC) estimate, given by Eq. (97),

$$\mathbb{L} = \left\langle \mathbf{L}^{(r)^{-1}} \mathbb{B}^{(r)} \right\rangle^{-1} \tag{170}$$

where $\mathbb{L}^{(r)}$ and $\mathbb{B}^{(r)}$ are the elastic stiffness and stress localization tensors, Eq. (82), of SR grain (r). Note that an ELSC calculation for the determination of $\mathbb{L}$ is implemented in the VPSC code at the beginning of each deformation increment. In this way, the texture changes are also accounted for in the determination of the stiffness of the polycrystalline material element.

Combining Eq. (169) with the viscoplastic constitutive relation coming from VPSC, Eq. (93), leads to,

$$\Delta\boldsymbol{\epsilon} = \mathbb{L}^{-1}\Delta\boldsymbol{\sigma} + \Delta t\dot{\boldsymbol{\epsilon}}_{vp}^{(px)}\left(\boldsymbol{\sigma}^t + \Delta\boldsymbol{\sigma}; q_i^t\right). \tag{171}$$

The residual $X(\Delta\boldsymbol{\sigma})$ can be defined as a (nonlinear) function of the stress increment $\Delta\boldsymbol{\sigma} = \boldsymbol{\sigma}^{t+\Delta t} - \boldsymbol{\sigma}^t$, which the UMAT should return to the FEM, as the constitutive response of the polycrystalline material point to the trial strain increment $\Delta\boldsymbol{\epsilon}^{FE}$

$$\mathbf{X}(\Delta\boldsymbol{\sigma}) = \Delta\boldsymbol{\epsilon} - \Delta\boldsymbol{\epsilon}^{FE} = \mathbb{L}^{-1}\Delta\boldsymbol{\sigma} + \Delta t\dot{\boldsymbol{\epsilon}}_{vp}^{(px)}\left(\boldsymbol{\sigma}^t + \Delta\boldsymbol{\sigma}; q_i^t\right) - \Delta\boldsymbol{\epsilon}^{FE} \tag{172}$$

The condition $\mathbf{X}(\Delta\boldsymbol{\sigma}) = 0$ (i.e., $\Delta\boldsymbol{\epsilon} = \Delta\boldsymbol{\epsilon}^{FE}$) is obtained using a Newton–Raphson (NR) scheme to solve this nonlinear algebraic equation. The corresponding Jacobian, $\mathbb{J}_{NR}$, is given by:

$$\frac{\partial \mathbf{X}(\Delta\boldsymbol{\sigma})}{\partial(\Delta\boldsymbol{\sigma})} = \mathbb{J}_{NR}(\Delta\boldsymbol{\sigma}) = \mathbb{L}^{-1} + \Delta t \frac{\partial \dot{\boldsymbol{\epsilon}}_{vp}^{(px)}}{\partial(\Delta\boldsymbol{\sigma})}\left(\boldsymbol{\sigma}^t + \Delta\boldsymbol{\sigma}; q_i^t\right)$$
$$= \mathbb{L}^{-1} + \Delta t \mathbb{M}^{(px)}\left(\boldsymbol{\sigma}^t + \Delta\boldsymbol{\sigma}; q_i^t\right) \tag{173}$$

Hence, given a guess $\Delta\boldsymbol{\sigma}^{k-1}$ for the stress increment, the new guess is obtained as:

$$\Delta\boldsymbol{\sigma}^k = \Delta\boldsymbol{\sigma}^{k-1} - \mathbb{J}_{NR}^{-1}(\Delta\boldsymbol{\sigma}^{k-1})\mathbf{X}(\Delta\boldsymbol{\sigma}^{k-1}). \tag{174}$$

Moreover, using the first equalities of Eqs. (172) and (173),

$$\mathbb{J}_{NR} = \frac{\partial \mathbf{X}(\Delta\boldsymbol{\sigma})}{\partial(\Delta\boldsymbol{\sigma})} = \frac{\partial(\Delta\boldsymbol{\epsilon} - \Delta\boldsymbol{\epsilon}^{FE})}{\partial(\Delta\boldsymbol{\sigma})} = \frac{\partial(\Delta\boldsymbol{\epsilon})}{\partial(\Delta\boldsymbol{\sigma})} \tag{175}$$

the Jacobian matrix that the VPSC-based UMAT should pass to the FEM is obtained,

$$\mathbb{L}^{tg} = \frac{\partial(\Delta\boldsymbol{\sigma})}{\partial(\Delta\boldsymbol{\epsilon})} = \mathbb{J}_{NR}^{-1} = \left[\mathbb{C}^{-1} + \Delta t\, \mathbb{M}^{(px)}\right]^{-1}. \tag{176}$$

Eq. (176) provides a closed expression for the FEM Jacobian, as a function of the viscoplastic tangent moduli (which is calculated as part of the VPSC algorithm), the elastic stiffness of the aggregate, and the FE time increment. The use of this expression greatly reduces the overall computational cost because the polycrystal stress and the elasto-viscoplastic tangent stiffness tensor are obtained from the same calculation loop. Moreover, the use of Eq. (176) for the FEM Jacobian usually provides quadratic convergence to the macroscopic nonlinear equations.

Finally, it must be remarked that the above procedure is semiimplicit: implicit in the stress value, because internal equilibrium is checked at the end of the increment, but explicit in the internal variables q_i (such as grain orientations, morphology and hardening variables). The reason for the proposed algorithm to remain explicit in those internal variables is related to the difficulty and computational cost of deriving a residual and a Jacobian matrix that would contain literally thousands of internal variables. In any case, the explicit updating of the internal variables does not appear to change significantly the convergence of the macroscopic model in most cases, as shown by Segurado et al. (2012).

7. APPLICATION TO VIRTUAL TESTING OF POLYCRYSTALS

In order to demonstrate the potential of CHP, applications of the tools and strategies presented above are illustrated in this section, with examples of simulations of the mechanical behavior of polycrystals under different conditions.

7.1 Strength

The mechanical behavior of materials is usually characterized through a uniaxial stress–strain curves that contain the information about the most important mechanical parameters: elastic modulus, yield strength, strain hardening rate, and ductility. Thus, the prediction of uniaxial stress–strain curves is obviously the first step to validate the capabilities of CHP and there are many examples in the literature for different polycrystalline materials (see, among many others, Cruzado et al., 2015; Ghosh, Shahba, Tu, Huskins, & Schuster, 2016; Roters, Eisenlohr, Bieler, & Raabe, 2010; Roters, Eisenlohr, Hantcherli, et al., 2010; Shahba & Ghosh, 2016; Zhang et al., 2015). From a Materials Science perspective, the information required to carry out an accurate prediction of the stress–strain curve includes a detailed description of the microstructure, and of the hardening parameters that dictate the plastic deformation of the constituent single crystals, as well. While microstructural information can be *easily* obtained using modern characterization techniques, the quantification of hardening behavior has to be achieved using different strategies based on: (a) micromechanical tests, (b) inverse optimization, or (c) multiscale models. While multiscale approaches to predict the CRSSs of the different slip systems in engineering materials are not yet mature, one example of each of the other two strategies is detailed below.

7.1.1 Strength of IN718 Ni-Based Superalloy

Polycrystalline IN718 is a Ni–Fe-based superalloy widely used for structural applications up to 650–700°C because of its good castability and weldability, high strength and corrosion resistance. The microstructure of IN718 is made up by a Ni FCC solid solution which contains a dispersion of nm-sized γ' ($Ni_3(Al,Ti)$) and γ'' (Ni_3Nb) coherent precipitates within the grains together with μm-sized metal carbides and δ phase (Ni_3Nb) particles at grain boundaries (Fig. 5) (Radavich, 1989; Song, Lee, & Kim, 2005). The volume fractions of γ' and γ'' phases are in the range 3%–5% and 10%–20%, respectively,

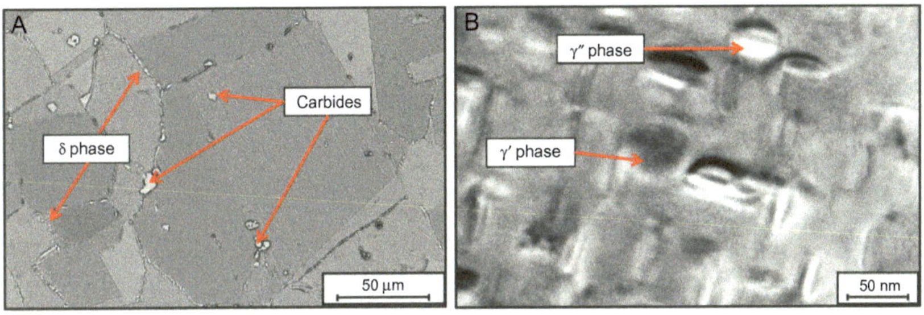

Fig. 5 Microstructure of IN718 Ni-based superalloy. (A) Polycrystal grain structure showing the distribution of metal carbides and δ phase at the grain boundaries and (B) distribution of γ' and γ'' precipitates within the Ni FCC solid solution. *Reprinted from Cruzado, A., Gan, B., Jimnez, M., Barba, D., Ostolaza, K., Linaza, A., ...* Segurado, J. (2015). Multiscale modeling of the mechanical behavior of IN718 superalloy based on micropillar compression and computational homogenization. Acta Materialia, 98, 242–253.

depending on the bulk alloy composition, the heat-treatment and the degree of element segregation (Fayman, 1987). The strength of the alloy in this case (and in the case of many other Ni-based superalloys) is provided by the interaction of the dislocations with the fine distribution of γ' and γ''. The precipitate size and spacing is of the order of 10–20 nm in wrought IN718 (Fig. 5B), which stands for the critical length scale that controls the mechanical behavior and, under these circumstances, micromechanical testing techniques that probe the mechanical response of the material in volumes of several μm^3 (much larger than the spacing between precipitates) can be used to determine the CRSS. Examples of these strategies in metallic alloys include micropillar compression (Cruzado et al., 2015) and nanoindentation (Hidalgo-Manrique et al., 2015; Patel & Kalidindi, 2017; Sánchez-Martín et al., 2014b; Vachhani & Kalidindi, 2015; Zambaldi, Zehnder, & Raabe, 2015).

The strategy to predict the mechanical behavior of polycrystalline Ni-based superalloys is depicted in Fig. 6 (Cruzado et al., 2015). The microstructural information included the grain size distribution obtained from optical microscopy images, that can be easily translated into the actual 3-D distribution in the case of equiaxed grains (Heilbronner & Bruhn, 1998), while the grain orientation was provided by X-ray diffraction measurements. The cubic RVE of the microstructure was generated from the Voronoi tessellation of a set of points, which was obtained from a Monte Carlo algorithm, so the grain size distribution in the RVE followed the experimental log-normal grain size distribution, following the strategy presented in Section 2.1. The grain orientations within the RVE were assigned to match the measured ODF.

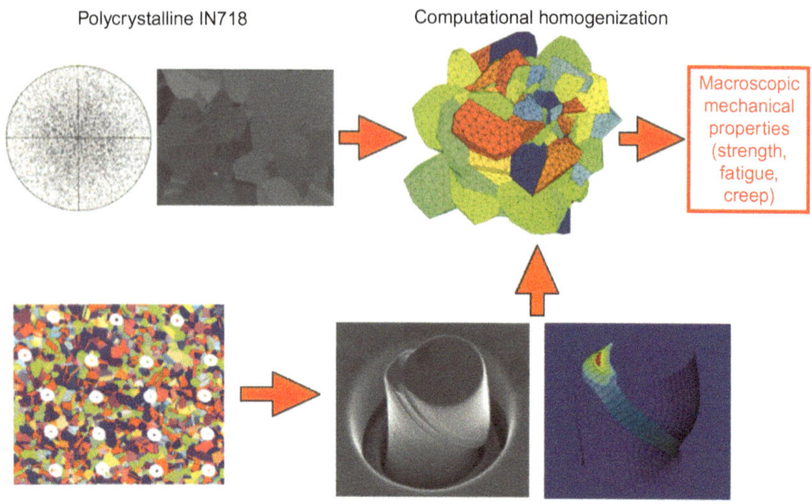

Fig. 6 Coupled experimental-simulation strategy to predict the mechanical properties of IN718 Ni-based superalloy using computational homogenization of polycrystals (Cruzado et al., 2015).

Deformation of the Ni solid solution occurred by dislocation slip along the 12 $\{111\}\langle 110\rangle$ slip systems of the FCC lattice. The crystal was assumed to behave as an elasto-viscoplastic solid in which the plastic slip rate in the slip system k, $\dot{\gamma}^k$, follows a phenomenological power-law dependency according to (Eq. 15)

$$\dot{\gamma}^k = \dot{\gamma}_0 \left| \frac{\tau^k}{g^k} \right|^{1/m} \mathrm{sgn}(\tau^k) \tag{177}$$

where $\dot{\gamma}_0$ stands for the reference strain rate, g^k is CRSS of the slip system k at the reference strain rate and m the strain rate-sensitivity parameter. The evolution of the CRSS of a given slip system, g^k, is expressed as,

$$\dot{g}^k = \sum_k h\, q_{kj} |\dot{\gamma}^j| \tag{178}$$

where h stands for the self-hardening modulus and q_{kj} are the interaction parameters that stand for the influence of hardening between different slip systems. The evolution of the self-hardening was described according to the Voce model,

$$h(\Gamma) = h_s + \left(h_0 - h_s + \frac{h_0 h_s \Gamma}{\tau_s - \tau_0} \right) \exp\left\{ \frac{-\Gamma h_0}{\tau_s - \tau_0} \right\} \tag{179}$$

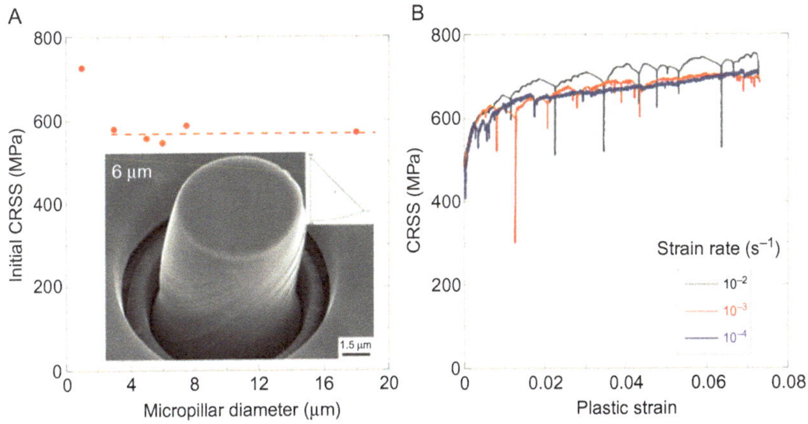

Fig. 7 (A) Initial CRSS as a function of the micropillar diameter for micropillars oriented for single slip along the ⟨235⟩ or ⟨245⟩ directions. One deformed micropillar showing the slip traces is depicted within the plot. (B) Effect of strain rate on the stress vs plastic strain curve in micropillars of 5 μm in diameter oriented for single slip. *From Cruzado, A., Gan, B., Jimnez, M., Barba, D., Ostolaza, K., Linaza, A., …* Segurado, J. (2015). *Multiscale modeling of the mechanical behavior of IN718 superalloy based on micropillar compression and computational homogenization. Acta Materialia, 98, 242–253.*

where h_0 is the initial hardening modulus, τ_0 the initial yield shear stress, τ_s the saturation yields shear stress, h_s the saturation hardening modulus at large strains and Γ stands for the accumulated shear strain in all slip systems, which is given by Eq. (18).

The parameters of this phenomenological constitutive equation were obtained from compression tests in single crystals carried out on micropillars with circular section extracted by means of focus ion beam milling from crystals oriented in different orientations. Initially, micropillars of different diameters oriented for single slip along the ⟨235⟩ or ⟨245⟩ directions were tested and the initial CRSS is plotted as a function of the micropillar diameter in Fig. 7A. It was found the initial CRSS (and the strain hardening rate, see Cruzado et al., 2015) were independent of the micropillar size (and, thus, representative of the bulk properties of the alloy) when the micropillar diameter was > 3 μm. Thus, micropillars of 5 μm in diameter were tested at different strain rates to determine the strain rate-sensitivity parameter m (Fig. 7B). In addition, micropillars oriented for double slip (coplanar and noncoplanar) as well as for multiple slip were tested. These experimental curves were used to determine the parameters of the constitutive model for IN718 (which are shown in Table 1) by comparison with finite element

Table 1 Parameters of the Crystal Plasticity Model for IN718 Ni-Based Superalloy at Ambient Temperature Obtained From Micropillar Compression Tests

$\dot{\gamma}_0 s^{-1}$	m	τ_0 (MPa)	τ_s (MPa)	h_0 (GPa)	h_s (GPa)	q
10^{-3}	0.017	466	599	6	0.3	1

Source: From Cruzado, A., Gan, B., Jimnez, M., Barba, D., Ostolaza, K., Linaza, A., ... Segurado, J. (2015). Multiscale modeling of the mechanical behavior of IN718 superalloy based on micropillar compression and computational homogenization. Acta Materialia, 98, 242–253.

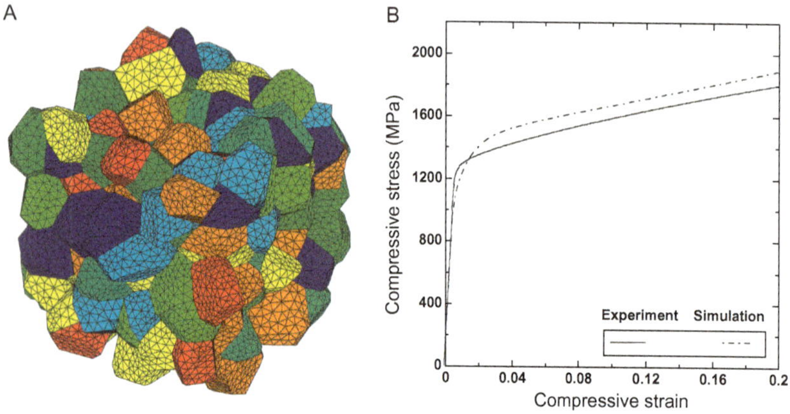

Fig. 8 (A) RVE of polycrystalline IN718 showing the finite element discretization. (B) Experimental and computational homogenization results of the true stress vs logarithmic strain curve in uniaxial compression of IN718 at ambient temperature. *From Cruzado, A., Gan, B., Jimnez, M., Barba, D., Ostolaza, K., Linaza, A., ...* Segurado, J. (2015). *Multiscale modeling of the mechanical behavior of IN718 superalloy based on micropillar compression and computational homogenization. Acta Materialia, 98, 242–253.*

simulations of the micropillar compression tests using the constitutive equation presented in Eqs. (177)–(179).

The RVE used to determine the mechanical behavior of polycrystalline IN718 by means of computational homogenization is depicted in Fig. 8A. It contained 210 grains and each grain discretized with at least 610 10-node quadratic tetrahedral elements. The grain size distribution and grain orientation followed the experimental data and simulations under periodic boundary conditions were carried out with four different realizations of the grain distribution. The differences among them were below 1.3% and the average true stress vs logarithmic strain curve under uniaxial compression obtained from computational homogenization is compared with the experimental data in Fig. 8B. The agreement between them was fairly good: the maximum difference in the compressive flow stress was below 4% and the prediction of the strain hardening rate was very accurate, validating the approach to predict the mechanical behavior of the alloy.

7.1.2 Strength of AZ31 Mg Alloys

Computational homogenization of HCP polycrystals to determine the strength presents several additional difficulties, as compared with cubic materials. First, HCP have different slip systems (basal, prismatic, pyramidal, etc.) with different slip resistances that have to be determined independently. Second, they often exhibit deformation twinning, which leads to very different hardening rates. Finally, most HCP metallic alloys present a very strong size effect during nanomechanical characterization (either using nanoindentation (Eswar Prasad, Rajesh, & Ramamurty, 2014; Nayyeri, Poole, Sinclair, & Zaefferer, 2017; Sánchez-Martín et al., 2014 b) or micropillar compression (Byer & Ramesh, 2013; Liu et al., 2017)). Thus, the actual values of the CRSS for slip and twinning in the bulk alloy cannot be obtained using the strategy presented in the previous section.

Under these circumstances, the best option to obtain the actual values of the CRSS for slip and twining is to use inverse optimization strategies, as the one depicted in Fig. 9 for AZ31 Mg alloy (Herrera-Solaz, LLorca, et al., 2014). The first step of the computational homogenization is the construction of an RVE which includes all the relevant microstructural details (grain size, shape, orientation) and a CP model that takes into account all the actual deformation mechanisms, namely basal, prismatic and pyramidal slip,

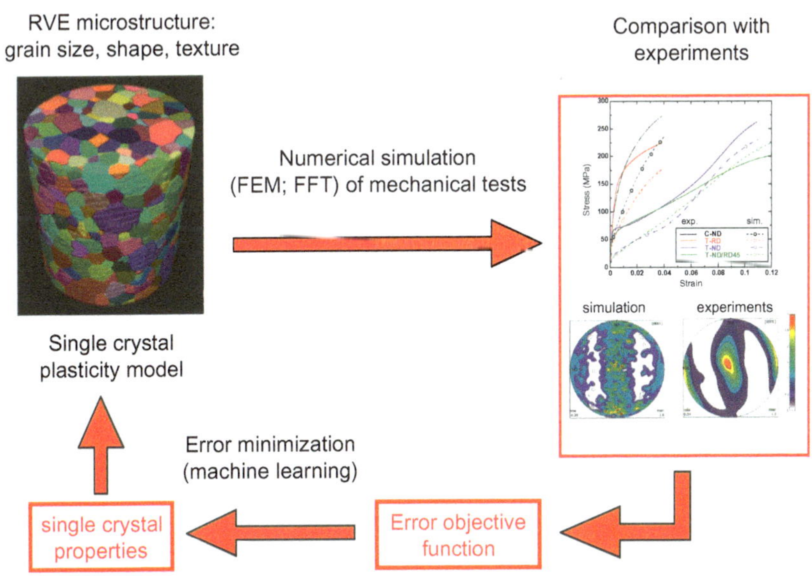

Fig. 9 Inverse optimization strategy to predict the mechanical properties of AZ31 Mg alloy using computational homogenization of polycrystals (Herrera-Solaz, LLorca, et al., 2014).

together with tensile twinning. The plastic slip rate for a given slip system k follows a power-law according to Eq. (15). Similarly, the twinning rate, $\dot{f}$, also follows a viscous law given by Eq. (55).

The activation of each deformation mode depends on the CRSS (g^k or g^α) for each slip or twinning system. The initial values (in absence of previous plastic deformation) of the CRSSs are given by τ_0^k or τ_0^α for the slip system k or the twin system α, respectively. A phenomenological hardening model was considered for the evolution of the CRSSs, which is able to reproduce the different stages of single crystal deformation (Kothary & Anand, 1998). The evolution of the CRSSs, g^k for slip and g^α for twinning, is expressed by Eqs. (180) and (181), respectively,

$$\dot{g}^k = q_{sl-sl}\sum_{j=1}^{3} h_0^j\left(1-\frac{\tau^j}{\tau_{sat}^j}\right)^{a_{sl}}|\dot{\gamma}^j| + q_{tw-sl}h_0^{tw}\left(1-\frac{\tau^\beta}{\tau_{sat}^{tw}}\right)^{a_{tw}}|\dot{\gamma}^\beta| \tag{180}$$

$$\dot{g}^\alpha = q_{tw-tw}h_0^{tw}\left(1-\frac{\tau^\alpha}{\tau_{sat}^{tw}}\right)^{a_{tw}}\dot{f}^\alpha\gamma_{tw} \tag{181}$$

where τ^j and τ^α stand for the resolved shear stress on the corresponding slip (either basal, prismatic, or pyramidal) and twinning system, $\dot{\gamma}^j$ is the plastic shear strain rate along system j, $\dot{f}^\alpha$ is the rate of the volume fraction transformation in the twin system α, and γ_{tw} is the characteristic shear of the twinning mode ($\gamma_{tw} = 0.129$ for extension twinning of Mg alloys, Herrera-Solaz, LLorca, et al. (2014)). It should be noted that extension twinning is a polar mechanism and it will only take place when the applied deformation leads to extension of the c axis of the HCP lattice.

The different parameters in these equations define the contributions arising from self-hardening and latent hardening. The self-hardening of a given slip (k) or twinning (α) system is defined by three terms: the saturation stress, τ_{sat}, the initial hardening rate h_0, and the hardening exponent a. The latent hardening contribution to slip due to slip in other systems is introduced with the coefficient q_{sl-sl} whereas the contribution induced by twinning is given by q_{tw-sl}. The model only takes into account the effect of twinning on slip and it is assumed that slip does not influence twinning ($q_{sl-tw} = 0$) following Capolungo, Beyerlein, and Tomé (2009) and Zhang and Joshi (2012).

One obvious difficulty is to find the actual values of the many parameters of the model. In most cases, they have been fitted by comparison with experimental results on polycrystals using a trial and error approach (Kalidindi et al., 2010; Miehe, Schotte, & Lambrecht, 2002; Miehe et al., 1999; Pei, Yun-Chang, & Qing, 2011; Segurado et al., 2012; Segurado & LLorca, 2013; Tomé, Agnew, & Yoo, 2001; Tomé, Lebensohn, & Necker, 2002; Tomé, Maudlin, Lebensohn, & Kaschner, 2001; Wu, Tomé, Agnew, Raeisinia, & Wang, 2010) but finding the optimum parameter set is neither easy nor a unique result is guaranteed. In fact, it is not unusual to find that different authors report different (or even contradictory) values of the CRSSs of slip and twinning for similar materials. To overcome this limitation, Herrera-Solaz, LLorca, et al. (2014) and Herrera-Solaz, Segurado, and LLorca (2015) proposed an optimization algorithm which is based in the construction of an objective error function $O(\boldsymbol{\beta})$ defined as

$$O(\boldsymbol{\beta}) = \sum_{i=1}^{p} |y_i - f(x_i, \boldsymbol{\beta})| = \|\mathbf{y} - \mathbf{f}(\boldsymbol{\beta})\|. \tag{182}$$

where x_i, y_i are a set of p points defining an experimental data set of the polycrystal (for instance, stress–strain curves in different orientations, pole figures after deformation, evolution of the fraction of twinned material, etc.) and $y_i = f(x_i;\boldsymbol{\beta})$ are the predictions of the corresponding data obtained by means of the numerical simulation of an RVE of the microstructure in which the mechanical behavior of each grain within the polycrystal is defined by a CP model, which depends on a set of parameters $\boldsymbol{\beta}$. The minimization of the error function can be carried out using different algorithms, such as Levenberg–Marquardt (Herrera-Solaz, LLorca, et al., 2014; Herrera-Solaz et al., 2015) or Nelder–Mead (Chakraborty & Eisenlohr, 2017; Zambaldi, Yang, Bieler, & Raabe, 2012). It should be noted that novel machine-learning strategies are very promising to carry out this optimization process (Mueller, Kusne, & Ramprasad, 2017) and also that the optimization strategy often requires several iteration loops. Thus, efficient numerical algorithms based on FFT will become critical within this framework.

The ability of the inverse optimization strategy to provide accurate values of the CRSS in complex HCP materials, like rolled AZ31Mg alloys processed by hot rolling, was demonstrated by Herrera-Solaz, LLorca, et al. (2014), Fig. 10A. The average grain size of the rolled AZ31 Mg alloy was 25 μm and the pole figure of the as-rolled material is plotted in Fig. 10B. It shows the strong basal texture typical of rolled Mg alloys, with the c axis parallel to the normal direction (ND). Also, the spread prismatic

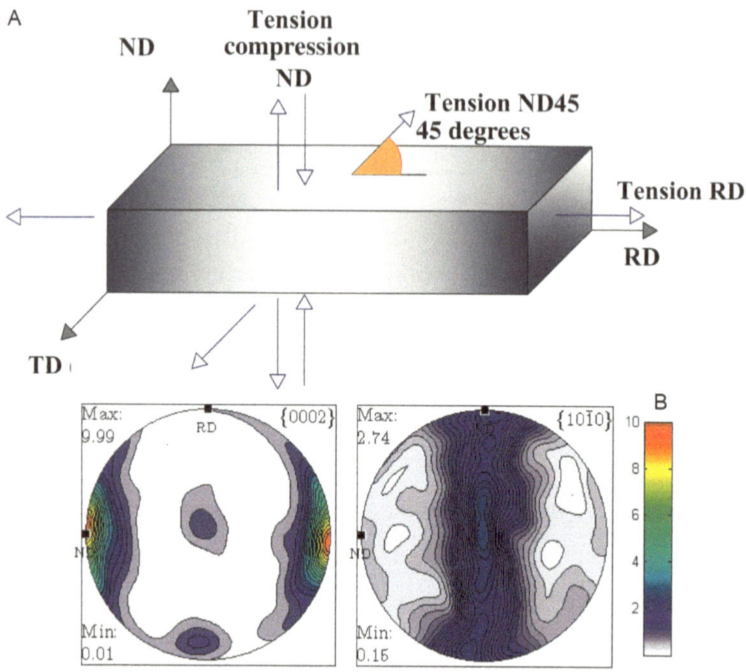

Fig. 10 (A) Slab of rolled AZ31 Mg alloy showing the RD, ND, and TD orientations as well as the orientation of the sample for the mechanical tests. (B) Pole figure of the rolled AZ31 alloy showing the strong basal texture. *Reprinted from Herrera-Solaz, V., LLorca, J., Dogan, E., Karaman, I., & Segurado, J. (2014). An inverse optimization strategy to determine single crystal mechanical behavior from polycrystal tests: Application to AZ31 Mg alloy. International Journal of Plasticity, 57, 1–15.*

poles appear along the rolling (RD) and transverse (TD) directions. Specimens for tension and compression experiments along different orientations indicated in Fig. 10A were machined from the plate and representative stress–strain curves of the mechanical tests can be found in Fig. 11. These curves show the strong plastic anisotropy of Mg alloys, which is triggered by the limited number of slip systems and by the polar nature of extension twinning, which is only activated when deformation leads to an extension of the *c* axis. As a result, deformation of wrought Mg alloys is markedly dependent on the orientation, and different slip systems (and in different order) are activated as a function of the loading direction (either tension or compression). For instance, tensile tests along ND show the typical S-shape of the stress–strain curve characteristic of twinning-dominated deformation, while the compression test along ND was dominated by pyramidal slip.

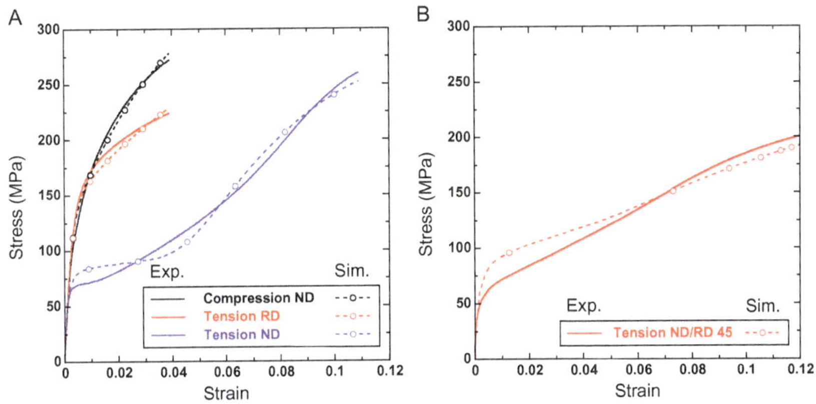

Fig. 11 (A) Experimental (*solid lines*) and numerical (*broken lines with symbol*) stress–strain curves resulting from the optimization procedure. (B) Model prediction of the tensile test in the RD-ND plane at 45 degrees from both orientations. *Solid lines* correspond to experimental results, while *broken lines with symbols* stand for the numerical simulations. *Reprinted from Herrera-Solaz, V., LLorca, J., Dogan, E., Karaman, I., & Segurado, J. (2014). An inverse optimization strategy to determine single crystal mechanical behavior from polycrystal tests: Application to AZ31 Mg alloy. International Journal of Plasticity, 57, 1–15.*

The stress–strain curves corresponding to the tests along ND in tension and compression and in tension along RD were used to determine the parameters of the CP model using the Levenberg–Marquardt method. Full-field homogenization was carried out on a RVE of the microstructure discretized with cubic finite elements under periodic boundary conditions. The RVE contained 584 grains and was created with the microstructure generator (Dream.3D, 2018). On average, each grain was discretized with seven voxels. The grains were equiaxed and the grain size followed a log-normal distribution with an average grain volume equal to the RVE divided by 584. The experimental stress–strain curves in the three different cases used to calibrate the constitutive equation (tension and compression along ND and tension along RD) were in very good agreement with the simulation predictions, as shown in Fig. 11A. Moreover, this set of parameters was used to predict the stress–strain behavior of the specimens tested in tension at 45 degrees from the ND/RD orientations (Fig. 11B). The agreement was again good and demonstrates the capabilities of inverse optimization approach to perform virtual test of polycrystals.

Further investigations (Herrera-Solaz et al., 2015) were carried out to analyze the robustness of the Levenberg–Marquardt optimization strategy to obtain the CRSSs of the different slip modes and extension twinning in an AZ31 Mg alloy. It was found that the results were not influenced by the properties of the single crystals chosen as starting point for the optimization process in so far they were reasonable for the Mg alloy considered. Very similar values of the CRSSs were obtained regardless of whether the initial plastic behavior of the single crystals was isotropic, anisotropic, or highly anisotropic. In addition, it was found that there is a minimum critical information that has to be supplied to the optimization strategy in order to achieve meaningful results. In the case of AZ31 Mg alloy, plastic deformation in the basal plane is controlled by basal slip while plastic deformation along the *c* axis accommodated by either twinning or a combination of prismatic and pyramidal $\langle c + a \rangle$ slip. Thus, a minimum of two stress–strain curves (one in which deformation along the *c* is controlled by twinning and another in which is controlled by prismatic and pyramidal $\langle c + a \rangle$ slip) are necessary as input for the inverse optimization algorithm. Including a third input stress–strain curve in the optimization process improved the accuracy of the results from the quantitative viewpoint.

7.2 Fatigue Life

Fatigue crack nucleation in metallic alloys is a process that involves several length scales, including the features of dislocation slip and dislocation/dislocation interactions at the nm scale and the collective motion of dislocations and the formation of particular dislocation patterns at the microscale. They give rise to the localization of plastic deformation in slip bands within the grains that will be the origin of embryonic cracks (Suresh, 2012). This process is very dependent on the local details of the microstructure (grain size, twins, texture) as well as on the presence of defects (pores or intermetallic inclusions). Thus, the role of the microstructure is much more critical in fatigue that in the case of strength. In the latter, the yield strength and the strain hardening are mainly controlled by the average value of the microstructural features (as given by the average grain size, pole figure, etc.) and little differences are found in polycrystals with the same average features until localization of damage occurs at the last stages of deformation. On the contrary, fatigue crack nucleation occurs by the localization of damage and mainly depends on the extreme values of the statistical distribution of the microstructural features, rather than on the average values. This leads to a very large experimental scatter among "nominally" identical polycrystals (from the viewpoint of average features) and makes very difficult to establish a solid link between microstructure and properties.

Obviously, only simulation techniques that take into account the microstructural details are able to provide the link between microstructure and fatigue properties. This is the reason why microstructure-sensitive computational modeling of fatigue (McDowell & Dunne, 2010; Pineau, McDowell, Busso, & Antolovich, 2016) is indeed among the applications of CHP that has sparked more interest in the last decade because it provides a unique framework to understand the influence of the microstructure and of the defects on the fatigue behavior of engineering materials. Moreover, metrics can be develop to predict the fatigue life as a function of the loading conditions and of the microstructure of the polycrystal, providing an engineering tool to relate microstructure and fatigue performance. This opens the path to account for design allowables that are controlled by the inherent microstructural variability in the material and to design optimum microstructures for a given loading condition.

The modeling framework to predict the fatigue behavior of polycrystalline metallic alloys by means of CHP is depicted in Fig. 12. The first step is to

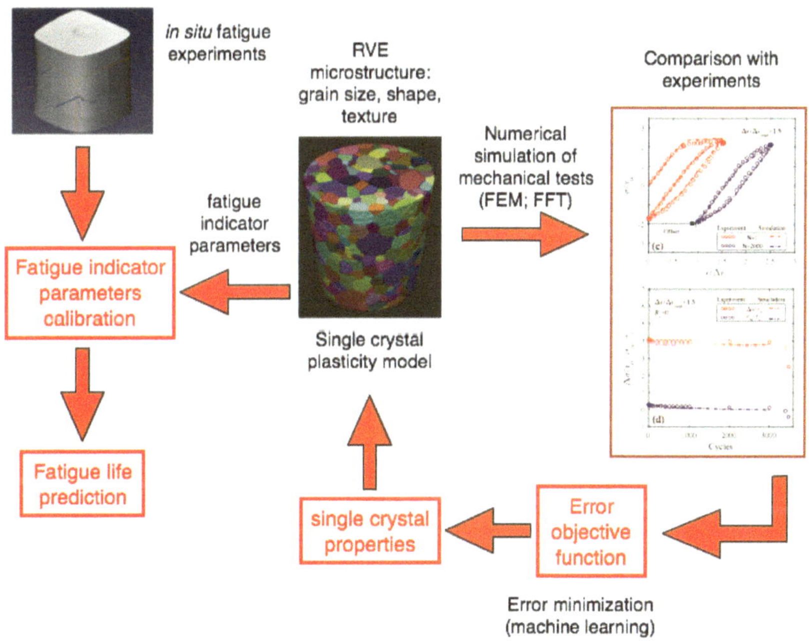

Fig. 12 Coupled experimental-simulation strategy to predict the fatigue life of polycrystalline metallic alloys using computational homogenization of polycrystals.

obtain the full micromechanical fields during cyclic deformation by means of the numerical simulation of the mechanical behavior of an RVE. The constitutive model for the single crystals should be able to account for the main features found during cyclic deformation of metallic alloys, namely Bauschinger effect, mean stress relaxation, ratcheting, and cyclic softening/hardening. This is normally achieved by means of phenomenological models (Section 3.1), whose parameters have to be calibrated for each alloy by comparison with the experimental cyclic stress–strain curves. This task can be done by means of trial and error approaches or optimization algorithms based in the construction of an objective error function (Cruzado et al., 2017).

One critical factor to obtain the parameters of the phenomenological CP model as well as to predict the fatigue life is that the mechanical response of the polycrystal changes with the number of cycles until a stable stress–strain hysteresis loop is attained. Stabilization of the mechanical behavior may take from few of cycles up to a few hundreds or thousands of cycles, depending of the alloy and of the loading conditions (Cruzado et al., 2017). In the latter scenario, the computational cost makes impossible the simulation of the mechanical response of the RVE and different cyclic jump strategies have been proposed to reach such large number of cycles without the explicit simulation of each of them (Ghosh & Dimiduk, 2011). Among them, Joseph, Chakraborty, and Ghosh (2010) proposed a novel wavelet transformation-based multitime scaling (WATMUS) algorithm which allows to reduce the number of simulated fatigue cycles by a factor of up to 100. Other simpler approaches can also be found on the literature. For instance, Cruzado et al. (2017) proposed a methodology, following similar approaches in the context of fretting fatigue (McColl, Ding, & Leen, 2004), based on the linear extrapolation of the evolution of the internal variables at the constitutive equation level, followed by one or several stabilization cycles.

Once the stabilized cyclic stress–strain hysteresis loop has been determined by mean of computational homogenization of the behavior of the RVE, the prediction of the fatigue life is based on some fatigue indicator parameter (FIP) obtained from the evolution of mechanical fields and internal variables at the local level within the RVE. Different FIPs have been reported in the literature to describe the main driving force that controls crack formation and they can be linked, using phenomenological relations, to some stage of the fatigue life of the alloy under study (crack nucleation, small crack propagation, or full fatigue life).

The most common FIPs found in the literature are related to the accumulation of some variables during one fatigue cycle, such as the accumulated plastic strain per cycle, $P(\mathbf{x})$ (Manonukul & Dunne, 2004; McDowell, Gall, Horstemeyer, & Fan, 2003; Sweeney, McHugh, McGarry, & Leen, 2012; Sweeney, O'Brien, Dunne, McHugh, & Leen, 2015; Sweeney, O'Brien, McHugh, & Leen, 2014)

$$P(\mathbf{x}) = \int_{cyc} \sqrt{\frac{2}{3}\mathbf{L_p}(\mathbf{x}) : \mathbf{L_p}(\mathbf{x})}\, dt \tag{183}$$

or the maximum plastic shear strain $\Delta\gamma(\mathbf{x})$

$$\Delta\gamma(\mathbf{x}) = \max_k \left\{ \int_{cyc} |\dot{\gamma}^k(\mathbf{x})|\, dt \right\} \tag{184}$$

which are based exclusively on the local plastic strain fields. The Fatemi–Socie FIP, FS, is given by

$$FS(\mathbf{x}) = \max_k \left\{ \frac{\Delta\gamma^k(\mathbf{x})}{2} \left[1 + k\frac{\sigma_n^k(\mathbf{x})}{\sigma_y} \right] \right\} \tag{185}$$

where $\Delta\gamma^k$ and σ_n^k stand for the maximum plastic shear strain and the maximum normal stress perpendicular to the slip plane of the k slip system and σ_y is the yield stress. This parameter was originally develop to rationalize the fatigue behavior of laboratory specimens (Fatemi & Socie, 1988, 1989) but it was used as the driving force to nucleate a crack in both low cycle fatigue (LCF) and high cycle fatigue (HCF) simulations based on CHP (Bennett & McDowell, 2003; Castelluccio & McDowell, 2013, 2014; Musinski & McDowell, 2012; Przybyla, Musinski, Castelluccio, & McDowell, 2013; Shenoy, Zhang, & McDowell, 2007).

Other popular FIP is the energy dissipated per cycle, $W(\mathbf{x})$, introduced by Charkaluk, Bignonnet, Constantinescu, and Dang Van (2002) and Korsunsky, Dini, Dunne, and Walsh (2007)

$$W(\mathbf{x}) = \max_k \left\{ \int_{cyc} \tau^k(\mathbf{x})\dot{\gamma}^k(\mathbf{x})\, dt \right\} \tag{186}$$

and the energy density stored in each cycle, $G(\mathbf{x})$ (Sweeney et al., 2012, 2015; Sweeney, O'Brien, McHugh, & Leen, 2014; Wan, Jiang, MacLachlan, & Dunne, 2016), which is expressed as

$$G(\mathbf{x}) = \int_{cyc} \xi \frac{\boldsymbol{\sigma}(\mathbf{x}) : \dot{\boldsymbol{\epsilon}}_p(\mathbf{x})}{\sqrt{\rho_{\mathrm{SSD}} + \rho_{\mathrm{GND}}}} dt \tag{187}$$

where ξ is a factor that stands for the fraction of the external work in a cycle that has been accumulated though the creation of dislocation structures. Its value is around 0.05 taking into account the typical fraction of the energy dissipated with respect to the external work (Wan, Jiang, et al., 2016).

Many other FIPs can be found in the literature (Anahid, Samal, & Ghosh, 2011; Ozturk, Shahba, & Ghosh, 2016; Sangid, Maier, & Sehitoglu, 2011) but they are not included here for the sake of brevity. There is no clear rule for selecting one of the different FIPs presented to predict the fatigue life and, indeed, different authors rely on one or another FIP for similar materials. The FIP that better correlates a fatigue process depends on the length scale considered, the characteristics of the alloy and also on the fatigue stage studied (incubation or small crack growth). Persistent slip bands form driven by the accumulation of plastic strain at the single crystal level and under uniaxial loading and both P and $\Delta\gamma$ are able to reproduce this observation (Manonukul & Dunne, 2004). The role of stress in the development of persistent slip bands and small cracks is much more important at the polycrystal level and under multiaxial loading and this is better accounted for using other FIPs including both plastic slip and stress. A critical assessment of the ability of the different FIPs to correlate crack initiation implies comparison with in situ microscopic experiments (Chen, Jiang, & Dunne, 2018) or with lower scale models (Sangid, 2013).

The FIPs defined in Eqs. (183)–(187) are local values computed at each point $\mathbf{x}$ of the RVE. This local value depends, however, on the details of the discretization and it is not representative of a fatigue damaged region (Sweeney et al., 2013). From a physical viewpoint, the FIP should be averaged over a region representative of the crack incubation zone, while the volume averaging is indeed necessary to avoid spurious stress concentrations and minimize mesh size effects from the computational perspective, as noted in Castelluccio and McDowell (2015). Three different averaging volumes are normally used, namely finite element, grain (Shenoy et al., 2007) or slip band (Castelluccio & McDowell, 2015). In the latter, the FIP is averaged on a slip band of constant thickness parallel to each slip plane k at $\mathbf{x}$. After averaging, the maximum value of the averaged FIP in the whole RVE (either the maximum averaged FIP in all the finite elements or in all the grains or in all the slip bands) is used to determine the fatigue life as indicated below

(Castelluccio & McDowell, 2013; Cruzado, Lucarini, LLorca, & Segurado, 2018b; Shenoy et al., 2007).

The last ingredient to predict the fatigue life in the microstructure-sensitive computational fatigue models is to establish the link between the FIPs and the different stages of the fatigue life. In this respect, micromechanics-based fatigue models can be roughly classified in two groups. In the first one, different relationships are established between the FIPs and the fatigue crack nucleation and propagation stages. In the second one, it is assumed that most of the fatigue life is spent in the nucleation of a crack and a direct relationship is established between the FIPs and the fatigue life.

The first approach was introduced by McDowell et al. (2003) and considers at least three stages in the fatigue life: crack incubation, microstructurally small crack growth (MSC) and long crack growth (LC). The fatigue life N is given by

$$N = N_N + N_{\mathrm{MSC}} + N_{\mathrm{LC}}. \tag{188}$$

where N_N is the number of fatigue cycles to nucleate a crack at a favorable site in the microstructure whose length is of the order of some microstructural length scale such as the grain size. N_{MSC} stands for the number of cycles spent from the nucleation of the crack until it reaches a length of typically 3–10 times the grain size. Crack nucleation and propagation in these two stages are strongly dependent on the microstructure (grain size, texture, size, shape and spatial distribution of second phases, etc.) (Suresh, 2012). Finally, N_{LC} is the number of the cycles associated with the propagation of a crack which is much longer than the longest critical microstructural length scale (typically the grain size in polycrystals). This stage is not very dependent on the microstructure and can be accounted for with the classical crack propagation theories based on fracture mechanics. Thus, CHP is used in this approach to determine N_N and N_{MSC}. Different FIPs and different relationships between each FIP and either N_N or N_{MSC} can be established to take into account that the driving forces for both process do not have to be equivalent.

The physical process of crack incubation is not yet fully understood. Thus, the determination of the principal driving forces and the development of robust initiation criteria based on these forces is based on more or less phenomenological approximations. Plastic deformation is accumulated progressively during cyclic deformation because of irreversible slip leading to the formation of low-energy dislocation arrangements to accommodate the irreversible slip. This results in the strain localization in a small region and in the formation of the so-called persistent slip bands that are the precursors to crack initiation. The first relation to link a FIP to the cycles for crack nucleation based on dislocation mechanics was proposed by Tanaka and Mura (1981), and it has the form

$$N_N(\mathrm{FIP})^2 = \frac{K_{\mathrm{TM}}}{d_g} \tag{189}$$

where the FIP in the original work (Tanaka & Mura, 1981) was the accumulated plastic strain per cycle, Eq. (183), d_g stands for the grain size and K_{TM} is a material constant. Nevertheless, many other expressions and models can be found in the literature (Sangid, 2013) and there is no a clear consensus on the form and of variables that should be included. In most of the models in which the FIPs are explicitly related to N_N (Castelluccio & McDowell, 2013, 2014, 2015; Przybyla & McDowell, 2010, 2012; Przybyla et al., 2013; Shenoy, 2006; Shenoy et al., 2007), this process is linked with the FIPs related to the plastic strain (P and $\Delta\gamma$) following expressions based on the Tanaka and Mura model.

Microstructurally small crack growth is treated within the framework of CHP following traditional expressions of macroscopic fatigue crack growth in which the cyclic crack tip displacement range ($\Delta CTOD$) is used as driving force instead of the stress intensity factor range (McDowell et al., 2003). However, $\Delta CTOD$ is difficult to obtain in CHP because it would require to define a crack and a fine mesh around it (Tucker, Cerrone, Ingraffea, & Rollett, 2015) but FIPs computed in standard computational homogenization simulations (without the crack) can be used as an effective surrogate measure of $\Delta CTOD$. Therefore, the microstructural crack growth rate can be obtained as (Shenoy et al., 2007)

$$\frac{da}{dN} = K_{CTOD}\tau_\gamma\, FS\, a \tag{190}$$

where a is the crack length, τ_γ an averaged value of the critical resolved shear stresses in all the slip systems, g^k, FS the Fatemi–Socie FIP, Eq. (185), and K_{CTOD} a constant. Similar expressions but including a threshold have been used in Przybyla et al. (2013). Eq. (190) can be integrated from an initial crack length of one grain up to a length in which the crack becomes long with respect to the microstructure. This modeling approach has been widely used for modeling HCF in Ni-based superalloys (Castelluccio & McDowell, 2013, 2014; Musinski & McDowell, 2012; Przybyla et al., 2013; Shenoy et al., 2007), Al alloys (McDowell et al., 2003), and Ti alloys (Przybyla & McDowell, 2012).

The constant that appears in the models for crack nucleation or short crack propagation presented above has always to be obtained by comparison with some experimental data. This calibration of the model is, however, problematic because it is very difficult to determine experimentally the actual number of fatigue cycles to nucleate a crack or to propagate the crack in the small crack regime. While this may be feasible by means on in situ fatigue tests coupled with diffraction and phase-contrast X-ray microtomography (Birosca et al., 2009; Herbig et al., 2011), they are extremely expensive and time-consuming and the available experimental data are limited to macroscopic fatigue tests in which the crack nucleation event cannot be detected.

A simple approach to overcome this problem was introduced by Manonukul and Dunne (2004) assuming that the fatigue life is mainly dominated by crack nucleation. Thus, the number of cycles necessary for crack nucleation is identified with the total fatigue life and the relationship is established between the FIP and the fatigue life. Manonukul and Dunne (2004) also showed that this relationship was characteristic of the polycrystalline C263 Ni-based superalloy and independent of the loading conditions, etc. This direct relationship between the FIP and the fatigue life has been profusely used since then using different FIPs and also considering different expressions to relate the FIP with the fatigue life. For instance, FIPs based on accumulated plastic strain have been used to assess the effect of elastic anisotropy and other factors in the fatigue life of a ferritic steel (Sweeney et al., 2013). The cyclic strain energy dissipation FIP, W (Eq. 186) was used to study the behavior under LCF in Inconel 718 superalloy (Cruzado, Lucarini, LLorca, & Segurado, 2018a; Cruzado et al., 2018b), showing that a microstructure-based fatigue model was able to predict the bilinear Coffin–Manson regime observed experimentally in this alloy. The same FIP was used to predict the fatigue life of stainless steel medical devices (Sweeney et al., 2012) and of a CoCr alloy (Sweeney, O'Brien, McHugh, & Leen, 2014). The stored energy FIP G (Eq. 187) was used within a physically based SGCP framework to predict the fatigue life in a ferritic steel in (Wan, Jiang, et al., 2016) and the same modeling approach was also employed in Sweeney, O'Brien, Dunne, McHugh, and Leen (2014) to study fatigue in a CoCr alloy. The concept of stored energy as driving force for crack initiation has also been proposed by Sangid et al. (2011) for a Ni-based superalloy where the energy stored was computed using a model that includes continuum and atomistic effects.

As a final remark, it is important to note the statical nature of fatigue predictions based on CHP. The number of grains in the RVE is always much smaller than that in the actual microstructure and therefore the values of the

FIPs obtained in the simulation of different RVEs corresponding to the same microstructure can be quite different. Thus, in general, the results obtained with a single RVE are not fully representative of the fatigue life for a given microstructure and do not provide reliable information about the scatter resulting from the particular orientation arrangement. This limitation is often overcome with the concept of SRVEs (Jeulin et al., 2004) presented in Section 2. Fatigue estimations are usually obtained aggregating the results of many different RVEs which are statistically representative of the same microstructure (Castelluccio & McDowell, 2013; Shenoy, 2006).

7.2.1 Prediction of the Fatigue Life in IN718 Ni-Based Superalloy

To illustrate the fatigue life prediction process using CHP, an example of prediction of the fatigue life in IN718 Ni-based superalloy in the LCF regime at 400°C is presented next (Cruzado et al., 2017, 2018a, 2018b). The CP model used to represent the cyclic plastic behavior (Cruzado et al., 2017) included the mechanisms leading to Bauschinger effect, mean stress relaxation and cyclic softening and a summary of the constitutive equations was already presented in Eqs. (25)–(27). The fatigue life estimation was based on the plastic energy dissipated by cycle FIP, Eq. (186), averaged in slip bands (Castelluccio & McDowell, 2015). The resulting FIP of a particular RVE, was given by

$$W_{cyc} = \max_{i=1,nb} \left\{ \max_{k_i} \frac{1}{V_i} \int_{V_i} W_{cyc}^{k_i}(\mathbf{x}) dV_i \right\} \tag{191}$$

where k_i (= 1, 2, 3) corresponds to the three different slips systems contained in the slip plane parallel to the band i, V_i is the volume of that band and nb is the total number of bands in the microstructure, which is four times the number of elements in the RVE. The fatigue life estimation for this particular RVE (the number of cycles N_i) is obtained as function of W_{cyc} using a power-law relation, introduced by Cruzado et al. (2018b), to account for the change in deformation mechanism that controls the nucleation of fatigue cracks in this material: from localized deformation in a few grains at small cyclic strain ranges to homogeneous plastic deformation at large cyclic strain ranges. Thus,

$$N_i = \frac{W_{crit}}{(W_{cyc})^m} \tag{192}$$

where the parameters W_{crit} and m were obtained from two independent fatigue experiments corresponding to two different cyclic strain ranges

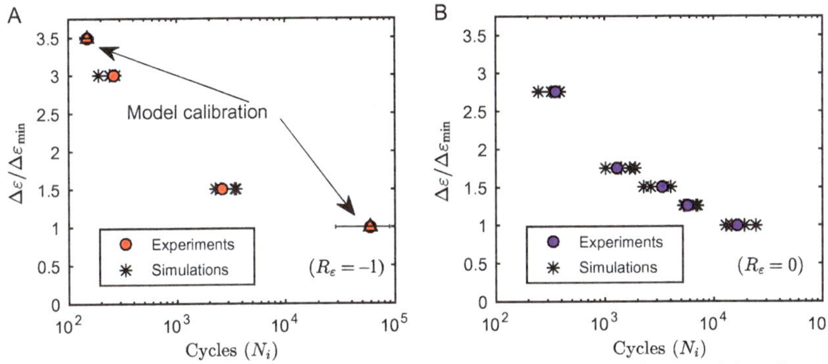

Fig. 13 Experimental results and model predictions for the fatigue life of IN718 Ni-based superalloy at 400°C. (A) $R_\varepsilon = -1$ and (B) $R_\varepsilon = 0$. The cyclic strain amplitude, $\Delta\epsilon$, is normalized by $\Delta\epsilon_{min}$, the minimum cyclic strain amplitude used in the tests. *Reprinted from Cruzado, A., Lucarini, S., LLorca, J., & Segurado, J. (2018b). Microstructure-based fatigue life model of metallic alloys with bilinear Coffin-Manson behavior. International Journal of Fatigue, 107, 40–48.*

under the same strain ratio $R_\varepsilon = -1$ (Fig. 13A). To this end, the FIPs (W_{cyc}) obtained by averaging the results of 20 different RVEs for each of the two loading conditions selected were used.

Finally, the model was used to estimate the fatigue life under different loading conditions. Predictions for each loading condition were obtained by averaging the FIPs computed using different RVEs. The ability of the proposed model to predict the fatigue life of IN718 alloy at 400°C is depicted in Fig. 13A and B, in which the experimental and the model results for the fatigue life are plotted as a function of the cyclic strain range in tests carried out with $R_\varepsilon = -1$ and 0, respectively. The model was able to predict accurately the fatigue life under small and large cyclic strain ranges for all the loading cases considered. Moreover, the numerical predictions with different RVEs led to estimations of the scatter in fatigue life which followed the same trends that the experimental observations: the scatter decreased as the applied cyclic strain range increased.

7.3 Void Growth in Polycrystals

Ductile failure of polycrystalline metals takes place by the nucleation, growth and coalescence of voids (Thomason, 1990). Void nucleation is normally triggered by either fracture or interface decohesion of second phase particles while void growth up to coalescence—that controls the final ductility of the material—is driven by the plastic flow and mainly depends on the stress state and on microstructural details. While the effect of stress state on void growth and coalescence has been analyzed in detail within the framework of isotropic plasticity models (Benzerga & Leblond, 2010), the influence of the microstructure (grain size, orientation) is far less understood. Nevertheless, it has been experimentally established that grain anisotropy and grain boundaries play an important role on void growth, particularly for low symmetry crystals (Lecarme et al., 2014; Nemcko & Wilkinson, 2016; Pushkareva et al., 2016).

CHP has been applied to analyze the effect of the microstructural factors on void growth using the dilatational viscoplastic FFT-based model described at the end of Section 5.2. This approach was used to simulate void growth in damaged polycrystalline materials under high stress-triaxiality loading that resembles that of typical Cu spall experiments (Escobedo et al., 2011; Escobedo, Cerreta, Dennis-Koller, Trujillo, & Bronkhorst, 2013). The simulation results in two cases in which the matrix behavior was different (FCC polycrystal vs isotropic matrix) were compared with postmortem orientation images from one of such experiments, showing grain orientations and voids in the region of incipient damage. This provides information on the effect of the matrix's polycrystallinity on porosity evolution, and identify microstructural effects on void growth, such as the influence of the Taylor factor of the crystalline ligaments that link interacting voids.

The mechanical behavior of the single crystals in the polycrystal assumed an FCC material with 12 $\{111\}$ $\langle 110 \rangle$ slip systems, a strain rate-sensitivity exponent $n = 5$ and a reference strain rate $\dot{\gamma}_0 = 1$, Eq. (15). The initial critical resolved shear stress was 100 MPa with a simple linear hardening $h_{kj} = 30$ MPa, Eq. (17). These values are consistent with the ones measured in shock-compressed Cu with 5% prestrain (Sencer, Maloy, & Gray, 2005). Meanwhile, for consistency, the constitutive parameters that describe the behavior of an analogous isotropic matrix material, Eq. (148), were $n = 5$, $\dot{\gamma}_0 = 1$, $\sigma_0 = 250$ MPa, and $H = 75$ MPa.

The polycrystalline RVE included 200 grains with an initial porosity of 1% distributed in 50 spherical voids randomly seeded at grain boundaries. It represents the spall region of a polycrystalline high-purity Cu target following compressive shock hardening and subsequent random nucleation of intergranular voids. The isotropic matrix RVE was built keeping the same void distribution and replacing the grains by a homogeneous isotropic matrix with the analogous constitutive behavior given above.

The initial porosity represented spallation voids that nucleate when interacting rarefaction waves produce the required tensile state for damage

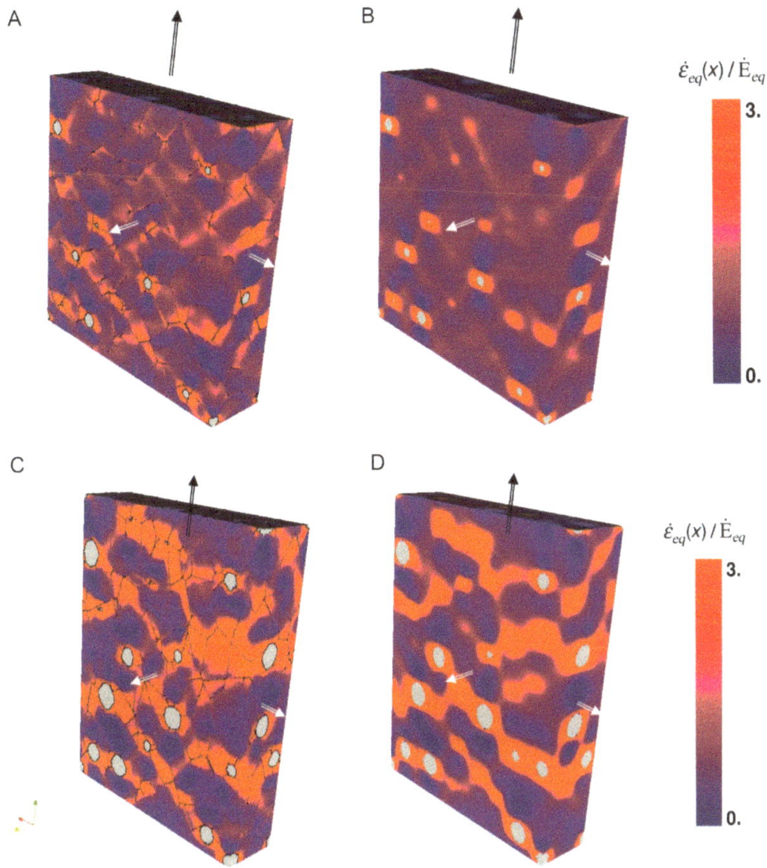

Fig. 14 Contour plots of the equivalent strain rate fields during spalling of a porous metal when the matrix is modeled as an FCC polycrystal with intergranular cavities (left), or a homogeneous isotropic matrix (right), for $X_\Sigma = 2.5$. (A) and (B) Initial, (C) and (D) final configurations. The *black arrow* indicates the direction of the largest principal stress, normal to the spall plane. Simulations were carried out using full-field homogenization with FFT. *Reprinted from Lebensohn, R. A., Escobedo, J. P., Cerreta, E. K., Dennis-Koller, D., Bronkhorst, C. A., & Bingert, J. (2013). Modeling void growth in polycrystalline materials. Acta Materialia, 61, 6918–6932.*

initiation. From that point on, it was assumed that the material deforms under an axisymmetric stress state, with far-field stresses $\Sigma_{zz} > \Sigma_{xx} = \Sigma_{yy}$ (where z is the shock direction, normal to the spall plane), and a constant stress-triaxiality $X_\sigma = \Sigma_m/\Sigma_{eq}$, where $\Sigma_m = (\Sigma_{zz} + 2\Sigma_{xx})/3$, and $\Sigma_{eq} = \Sigma_{zz} - \Sigma_{xx}$.

The local fields and the porosity evolution for both the polycrystal and the isotropic matrix cases are plotted in Fig. 14 for $X_\Sigma = 2.5$. Fig. 14A and C stands for the relative equivalent strain rate fields calculated at the initial and final deformation steps, respectively, in the case of the polycrystalline FCC matrix. The corresponding results for the isotropic matrix are plotted in Fig. 14B and D. The black arrows indicate the direction of the largest principal stress, normal to the spall plane. The volumetric deformation is fully accommodated by void growth, resulting in a strong localization of (incompressible) plastic deformation in the material surrounding the cavities. This is revealed by the high values (relative to the macroscopic equivalent strain rate) of the plastic strain rate fields, especially between cavities that are close to each other and whose ligaments are relatively well aligned with the direction of largest principal stress.

In order to better appreciate the differences between both cases, 2-D sections from the 3-D simulations of Fig. 14 are plotted in Fig. 15A–C for the porous material with isotropic matrix and in Fig. 15D–F the case of the FCC polycrystalline matrix at different levels of porosity. The experimental results of the typical postmortem EBSD-generated Taylor factor maps of a damaged polycrystalline Cu, from the spall region of the target are also shown in Fig. 15G and H (Escobedo et al., 2011, 2013). Void growth in the case of the isotropic matrix is only affected by the distance and relative position of the interacting voids. For example, the two pairs of voids within the yellow circle in Fig. 15C interact strongly and grow in a similar way. On the contrary, the voids in the lower pair (marked 2) do interact and grow profusely while the upper pair (marked 1) stop growing, likely due to the presence of a hard crystalline ligament, in the polycrystalline matrix case (Fig. 15F). Indeed, this ligament is mostly formed by grain 92 in this particular 2-D section, whose initial Taylor factor with respect to tension along the z direction is 3.31, i.e., a hard crystallographic orientation for deforming in tension along that axis. Interestingly, the EBSD Taylor factor maps show similar behavior. Fig. 15G shows two voids that are ideally located to interact, but they did not coalesce, likely because of the hard ligament linking them (note that the orange and red colors represent Taylor factors > 3.1). On the other hand, Fig. 15H shows three voids (the two on the right have already merged) in similar spatial configuration as those of Fig. 15G, but linked by soft orientations (Taylor factor <2.5), which have coalesced or are in an advanced state of precoalescence. It should be emphasized the qualitative character of the above comparison, which is based on two strong assumptions, namely: (1) that the effect of the 3-D neighborhood can be neglected, and (2) that the stress state remained axisymmetric throughout the entire deformation process, to justify the use of Taylor factors along the shock (axial) direction.

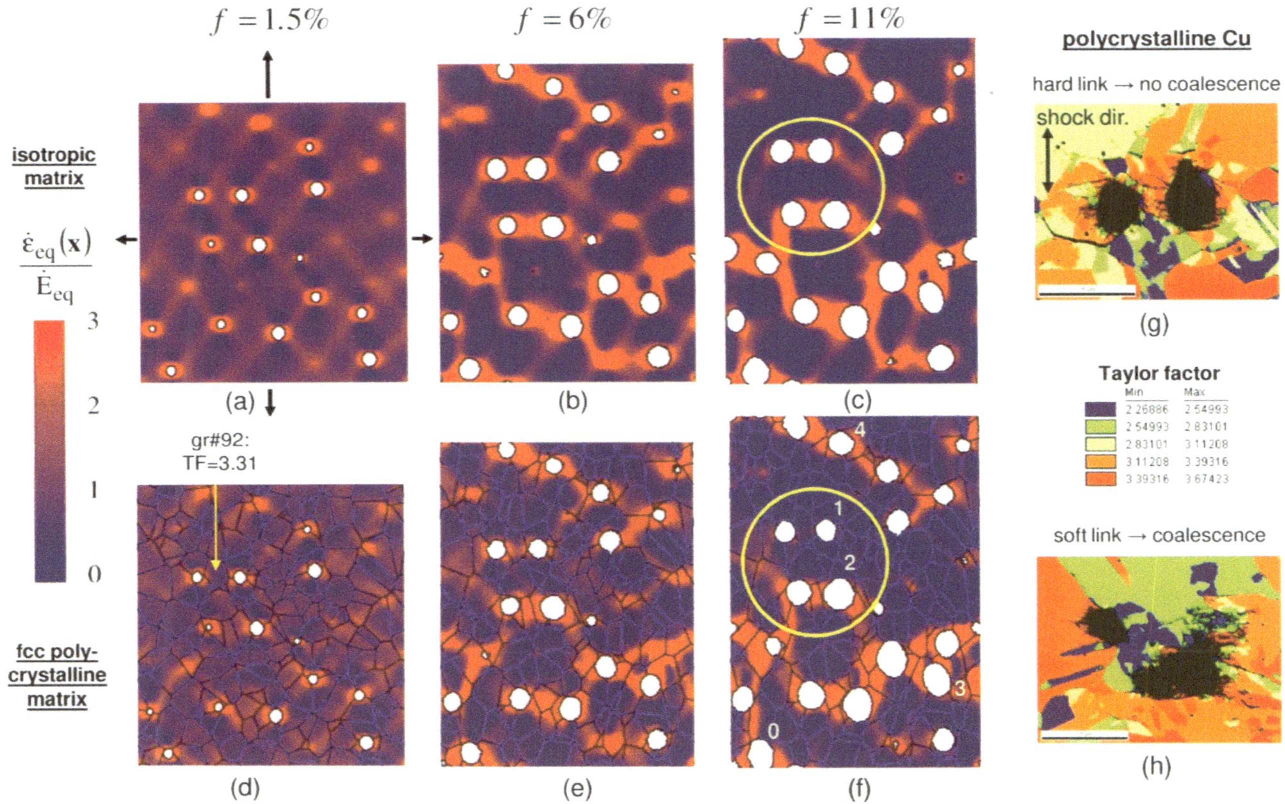

Fig. 15 2-D sections corresponding to the 3-D simulations in Fig. 14, at different levels of porosity, for porous materials with (A)–(C) isotropic matrix, and (D)–(F) FCC polycrystalline matrix. (G)–(H) Typical postmortem EBSD-generated Taylor factor maps of dynamically damaged polycrystalline Cu, recovered from the spall region. *Reprinted from Lebensohn, R. A., Escobedo, J. P., Cerreta, E. K., Dennis-Koller, D., Bronkhorst, C. A., & Bingert, J. (2013). Modeling void growth in polycrystalline materials. Acta Materialia, 61, 6918–6932.*

7.4 Multiscale Modeling of Rolling

Multiscale models with homogenization at the mesoscale (described in Section 6) are useful to simulate the mechanical behavior of polycrystals subjected to nonhomogeneous stress states. An example of application of this strategy is the analysis of the spatial variation of the texture evolution during rolling of an FCC polycrystal (Segurado et al., 2012). These variations, in turn, change the local behavior and, as a consequence, determine the redistribution of stresses during deformation. This case constitutes a challenge for the proposed approach, since it includes heterogeneous deformation, large strains and rotations, contact and complex local strain histories.

The simulation of rolling at the macroscopic level is carried out using the FEM. A parallelepipedic slab of material is rolled between two cylinders leading to 50% reduction in thickness (Fig. 16). The initial aspect ratio of the slab was 4×1 in the rolling-direction/normal-direction plane (RD-ND plane). Plane strain was imposed in the transverse direction, TD, and symmetry was used to reduce the size of the FEM model. The RD and TD coincided with the x and z axes of the macroscopic FEM. The slab was discretized with a structured mesh of linear cubes with full integration (C3D8 in Abaqus) using 26×9 elements in the RD-ND plane and 1 element in TD. Load was applied in two steps. The initial step consisted in a small vertical displacement of the cylinders in order to establish contact conditions between rolls and plate at the entry and the cylinders rotate at 2 rad/s in the second step until the plate has been completely rolled. The movement of the cylinders was transferred to the plate by friction, and the imposed angular velocity corresponds to an average compression strain rate component of along the normal direction (ND, y axis).

The mechanical behavior of the material at each Gauss point of the FEM was determined using the VPSC approach. The FCC polycrystalline material was initially represented by 500 randomly oriented grains which deform plastically in 12 $\{111\}\ \langle 110 \rangle$ slip systems. The elastic stiffness tensor of the Al FCC crystal was given by $L_{1111} = 108$ GPa, $L_{1212} = 62$ GPa, and $L_{4444} =$ 28 GPa. The VPSC constitutive parameters, adjusted to reproduce the experimental behavior of a 6116 Al alloy deformed in uniaxial compression (Tomé et al., 2002) followed a Voce law, Eq. (179), characterized by τ_o 116 MPa, $\tau_s = 119$ MPa, $h_o = 793$ MPa, and $h_s = 31$ MPa, isotropic hardening ($q_{kj} = 1, \forall k, j$), and a strain rate-sensitivity exponent $n = 10$.

The geometry of the rolled plate in the final stage is shown in Fig. 16, together with the corresponding contour plots of the Von Mises stress, the diagonal strain component in ND, ϵ_{yy}, and the shear strain, ϵ_{xy}. Interestingly,

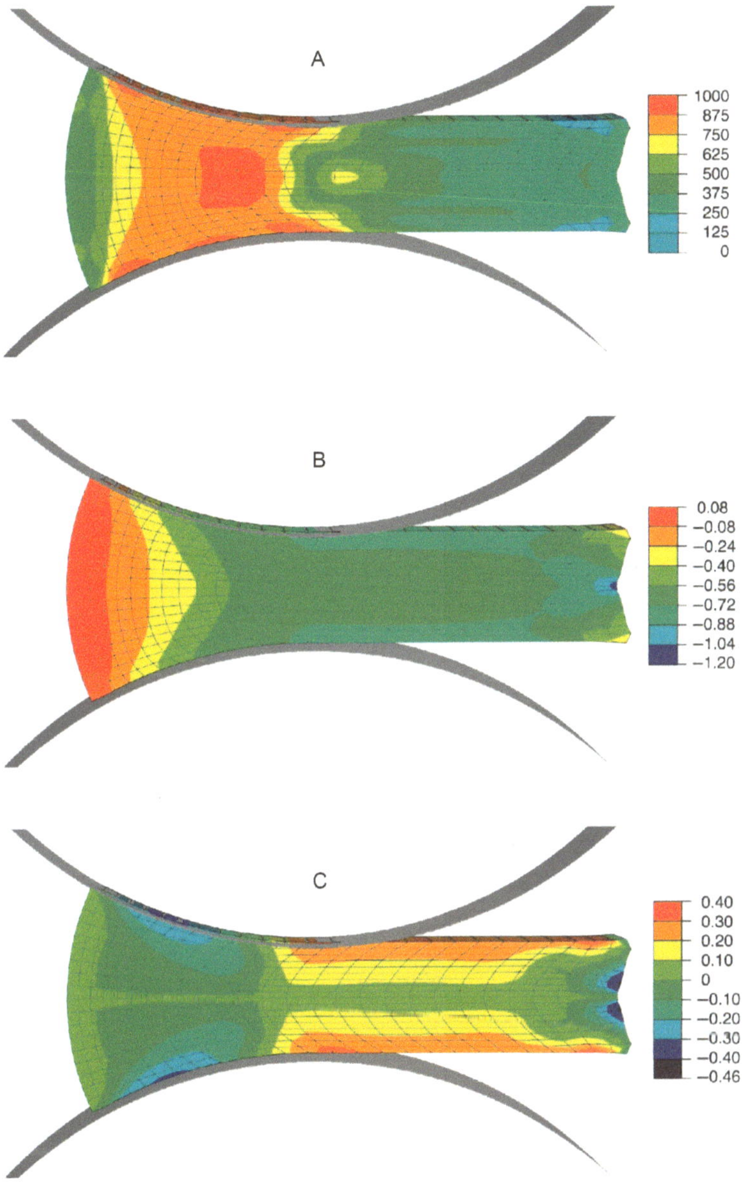

Fig. 16 Multiscale finite element simulation of rolling of an FCC plate, up to 50% reduction. Contour plots of: (A) Von Mises stress (in MPa). (B) Diagonal strain component in ND, ϵ_{yy} and (C) Shear strain component along RD normal to ND, ϵ_{xy}. *Reprinted from Segurado, J., Lebensohn, R. A., Llorca, J., & Tomé, C. (2012). Multiscale modeling of plasticity based on embedding the viscoplastic self-consistent formulation in implicit finite elements. International Journal of Plasticity, 28, 124–140.*

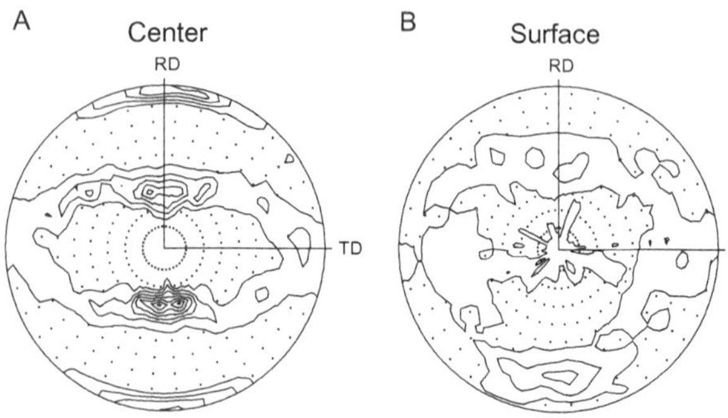

Fig. 17 {111} pole figures predicted by multiscale modeling strategy for two polycrystalline material points located: (A) near the center and (B) near the surface of the slab. *Lines* represent multiple of random distributions (mrd) and *dots* represent regions with less than 1 mrd. *Reprinted from Segurado, J., Lebensohn, R. A., Llorca, J., & Tomé, C. (2012). Multiscale modeling of plasticity based on embedding the viscoplastic self-consistent formulation in implicit finite elements. International Journal of Plasticity, 28, 124–140.*

the stress does not vanish in the region already rolled and a heterogeneous distribution of residual stresses is left in the material. It can be concluded from both Fig. 16C and the final shape of the elements that elements near the surface underwent significant shear deformation along RD normal to ND, while elements toward the center were subjected to smaller shear distortion. This shear deformation is a consequence of the friction between the metal surface and the cylinders.

The different deformation histories of the polycrystalline material points through the section determine different evolution of the texture. This can be appreciated in Fig. 17A and B, which show the $\langle 111 \rangle$ pole figures of two material points located near the center and near the surface of the plate, respectively. Both points were taken far from the leading tip in the rolling direction in order to minimize edge effects and obtain results that resemble those of a continuum rolling process. The textures near the center resemble typical plane strain compression textures of high stacking fault energy FCC polycrystals, while the texture associated with a point near the surface shows significant rotation with respect to TD of the rolling texture components, consistent with a plane strain + shear strain history applied to the material point.

8. CONCLUDING REMARKS AND FUTURE DEVELOPMENTS

CHP by means of mean-field approximations or full-field simulations of a RVE of the microstructure has become a mature tool in the last 20 years, enabling the simulation of the mechanical behavior of polycrystalline aggregates taking into account the effect of microstructure. This "micromechanical" approach is significantly different from the classical phenomenological plasticity models for polycrystals (for example, J_2 plasticity and associated damage and failure models). These models are known to provide an accurate response for the homogenized behavior at the macroscopic level and the parameters are calibrated by means of mechanical tests of polycrystalline specimens. Nevertheless, this classical methodology presents several limitations:

- Each set of parameters is only valid for a given set of testing conditions (temperature, strain rate, loading mode) and microstructural features (grain size distribution and morphology, texture). Any change in these factors requires the recalibration of the model.
- These phenomenological plasticity models are accurate to predict the homogenized properties (average behavior of many crystals), but fail to determine the macroscopic properties when they are controlled by microstructural features (fatigue and creep crack initiation, crack propagation, effect of defects, etc.).
- Phenomenological plasticity models assume that the material is homogeneous and do not include information about the microstructural details, such as grain size and orientation, intragranular distribution of metallurgical phases. Hence, they cannot be coupled with models dedicated to predict the formation and/or evolution of microstructure during thermomechanical processing.

Obviously, CHP has the potential to overcome these limitations by establishing a clear link between microstructure and mechanical properties that is unfeasible using classical plasticity models. Consequently, CHP is nowadays the standard research tool to analyze the mechanical behavior of polycrystalline aggregates and is starting to be adopted by industry to solve engineering problems (Roters, Eisenlohr, Bieler, & Raabe, 2010). This progress in industrial applications has been supported by the availability of different open-source software packages that can be used to represent the microstructure of polycrystals (Dream.3D, 2018; Neper, 2018) and to simulate the mechanical behavior using mean-field (Lebensohn & Tomé, 1993; Lebensohn et al., 2007) and/or full-field methods based on the FEM or the FFT (CAPSUL, 2018; Damask, 2018). The application of "virtual testing" and "virtual design" methodologies based on CHP is expected to have a significant industrial impact, as it has already happened in the field of composite materials for aerospace (González et al., 2017; LLorca et al., 2011). In particular, it will greatly reduce the number of experimental tests necessary to characterize the mechanical behavior of new materials and will provide guidelines to modify the microstructure of the latter in order to improve their mechanical performance under different loading conditions. In addition, the techniques used in the engineering design will not be based on huge and expensive experimental campaigns but on micromechanical models that can account for the variability in the microstructure.

Nevertheless, as CHP is used to tackle more and more complex situations, several important limitations of the current strategies to simulate the mechanical performance of polycrystals appear, which should be addressed in the future. The first important problem to be analyzed is the slip transfer across grain boundaries during deformation. There is ample experimental evidence showing that grain boundaries may act as strong barriers for dislocations, which are stopped at the boundary leading to the formation of pile-ups, or are easily transferred to the neighbor grain leading to the propagation of deformation on a different slip system (Bayerschen, McBride, Reddy, & Böhlke, 2016; Bieler, Eisenlohr, Zhang, Phukan, & Crimp, 2014; Hémery, Nizou, & Villechaise, 2018; Malyar, Micha, Dehm, & Kirchlechner, 2017). Similar effects have been found for twinning (Hong, Godfrey, & Liu, 2016; Kumar, Beyerlein, McCabe, & Tomé, 2016). The nature of the grain boundary from the viewpoint of slip transmission is assumed to depend on three angles that define the misorientation between the slip vectors, the slip plane normal and the slip plane intersections with the grain boundary plane. Different criteria have been proposed but the experimental evidence is still far from being conclusive (Bayerschen et al., 2016; Bieler et al., 2014). Nevertheless, the behavior of grain boundaries may be a determinant factor in the mechanical behavior of strongly textured polycrystals (where one type of grain boundary may be dominant) and in the nucleation of damage by shear band formation (Wang et al., 2014) or grain boundary cracking (Boehlert, Chen, Gutiérrez-Urrutia, Llorca, & Pérez-Prado, 2012; Cepeda-Jiménez, Molina-Aldareguia, & Pérez-Prado, 2015). This is particularly important in the case of low symmetry crystals, in which the limited availability of slip systems may led to large stress concentrations at the grain boundaries if slip transfer is hindered. Moreover, cracks are often nucleated at grain boundaries during fatigue and crack propagation across neighbor grains is critical to determine whether a microstructurally small crack will be arrested or propagate until failure (Musinski & McDowell, 2016; Wan, MacLachlan, & Dunne, 2016).

Grain boundaries are assumed to be transparent for slip transfer in standard (phenomenological) CP models and, thus, the effect of the grain boundary on the slip transmission is not accounted for. Physically based CP models can introduce the effect of grain boundaries through different parameters that take into account the increase in dislocation density near the grain boundary (Hauoala et al., 2018; Mayeur, Beyerlein, Bronkhorst, & Mourad, 2015; Su et al., 2016). Finally, lower-order SGCP account for the effect grain boundaries through the density of GNDs generated near the boundaries due to the deformation incompatibility between grains with different orientations, while higher order SGCP can control the dislocation flux across the boundary by means of the higher order boundary conditions. Thus, either physically based or SGCP models can be used to simulate the effect of grain boundaries on the plastic deformation of polycrystals and the development and experimental validation of these strategies will be an important area of research in the near future.

Another important development in CHP is the introduction of the "physical" twins in the simulations. So far, simulation of twinning during mechanical deformation has been taken into account through models that treat twinning as a pseudo-slip system (Abdolvand & Daymond, 2013a, 2013b; Chang & Kochmann, 2015; Kalidindi, 1998). Twinning is triggered by a CRSS on the twin plane and leads to a plastic deformation that depends on the volume fraction of twinned material and on the characteristic shear strain of the twin system considered. Moreover, plastic slip within the twinned regions can also be included. While this approach is accurate from the viewpoint of the effective properties of the polycrystal, it neglects the physical formation of a twinned region which is connected to the parent grain through two coherent grain boundaries. The importance of the presence of the twinned region in the deformation mechanisms has become very clear in the analysis of fatigue crack initiation in FCC metals, which show that fatigue cracks tend to nucleate in slip bands on $\{111\}$ planes parallel and, slightly offset from, annealing twins because of the stress concentration associated with the presence of the twin (Heinz & Neumann, 1990; Stinville, Lenthe, Miao, & Pollock, 2016; Stinville, Vanderesse, Bridier, Bocher, & Pollock, 2015). Obviously, this effect can only be accounted for if the "physical" twin is included within the grain in the numerical simulation. Annealing twins can be introduced into the initial RVE (although they require very fine meshing of the grains, leading to very large models) (Cerrone et al., 2015; Yeratapally, Glavicic, Hardy, & Sangid, 2016) but twins are often nucleated during deformation and should be introduced *on the fly* during the numerical simulation. This is necessary to account for the hardening effect of twinning on dislocation slip in a physical way and not through phenomenological models that cannot include the many different types of interactions. Strategies to achieve this goal have been presented recently (Ardeljan, McCabe, Beyerlein, & Knezevic, 2015; Cheng & Ghosh, 2017; Cheng, Shen, Mishra, & Ghosh, 2018; Liu et al., 2018) and they should be further developed.

Finally, the crystals in polycrystalline materials subjected to large plastic deformation do not only undergo a change in shape. Grain fragmentation and recrystallization processes lead to dramatic microstructural changes during deformation (particularly at high temperatures) that are used during thermomechanical processes to manufacture materials with new microstructures and properties. Coupling microstructure evolution with mechanical deformation will allow the "virtual processing" and "virtual design" of novel microstructures with optimized properties for specific applications and will dramatically increase the add value of CHP. Strategies to achieve these goals have been developed by coupling CHP with phase-field simulations (Abrivard, Busso, Forest, & Appolaire, 2012a, 2012b; Admal, Po, & Marian, 2018; Gvenc, Bambach, & Hirt, 2014; Lim et al., 2016; Vondrous et al., 2015) or cellular automata (Chuan, He, & Wei, 2013; Li, Sun, & Yang, 2016). They should be developed further to overcome the computational problems associated with the different time scales for mechanical deformation and microstructural evolution as well as the experimental and theoretical determination of the driving forces that control the microstructural evolution (grain boundary mobility, nucleation, etc.) under these complex conditions.

ACKNOWLEDGMENTS

This investigation was supported by the European Research Council under the European Unions Horizon 2020 research and innovation program (Advanced Grant VIRMETAL, grant agreement No. 669141), by the Spanish Ministry of Economy and Competitiveness through the project DPI2015-67667 and by Los Alamos National Laboratory's Laboratory-Directed Research and Development (LDRD) program.

REFERENCES

Abdolvand, H., & Daymond, M. R. (2013a). Multi-scale modeling and experimental study of twin inception and propagation in hexagonal close-packed materials using a crystal plasticity finite element approach. Part I: Average behavior. *Journal of the Mechanics and Physics of Solids*, *61*, 783–802.

Abdolvand, H., & Daymond, M. R. (2013b). Multi-scale modeling and experimental study of twin inception and propagation in hexagonal close-packed materials using a crystal plasticity finite element approach. Part II: Local behavior. *Journal of the Mechanics and Physics of Solids*, *61*, 803–818.

Abrivard, G., Busso, E., Forest, S., & Appolaire, B. A. (2012a). Phase field modelling of grain boundary motion driven by curvature and stored energy gradients. Part II: Application to recrystallisation. *Philosophical Magazine, 92*, 3643–3664.

Abrivard, G., Busso, E., Forest, S., & Appolaire, B. A. (2012b). Phase field modelling of grain boundary motion driven by curvature and stored energy gradients. Part I: Theory and numerical implementation. *Philosophical Magazine, 92*, 3618–3642.

Acharya, A., & Bassani, J. L. (1995). Incompatible lattice deformations and crystal plasticity. *Plastic and Fracture Instabilities in Materials, ASME Proceedings, 57*, 75–80.

Acharya, A., Bassani, J. L., & Beaudoin, A. (2003). Geometrically necessary dislocations, hardening, and a simple gradient theory of crystal plasticity. *Scripta Materialia, 48*, 167–172.

Admal, N. C., Po, G., & Marian, J. (2018). A unified framework for polycrystal plasticity with grain boundary evolution. *International Journal of Plasticity, 106*, 1–30.

Aifantis, E. C. (1987). The physics of plastic deformation. *International Journal of Plasticity, 3*, 211–247.

Alkemper, J., & Voorhees, P. W. (2001). Quantitative serial sectioning analysis. *Journal of Microscopy, 201*, 388–394.

Allison, J., Cowles, B., DeLoach, J., Pollock, T., & Spanos, G. (2013). *Integrated computational materials engineering (ICME): Implementing ICME in the aerospace, automotive, and maritime industries*. Warrendale, PA: The Minerals, Metals & Materials Society.

Allison, J., Li, M., & Wolverton, C. (2006). Virtual aluminum castings: An industrial application of ICME. *The Journal of The Minerals, Metals & Materials Society, 58*, 28–35.

Anahid, M., Samal, M. K., & Ghosh, S. (2011). Dwell fatigue crack nucleation model based on crystal plasticity finite element simulations of polycrystalline titanium alloys. *Journal of the Mechanics and Physics of Solids, 59*, 2157–2176.

Anand, L., & Kothari, M. (1996). A computational procedure for rate-independent crystal plasticity. *Journal of the Mechanics and Physics of Solids, 44*, 525–558.

Ardeljan, M., McCabe, R. J., Beyerlein, I. J., & Knezevic, M. (2015). Explicit incorporation of deformation twins into crystal plasticity finite element models. *Computer Methods in Applied Mechanics and Engineering, 295*, 396–413.

Arsenlis, A., & Parks, D. (1999). Crystallographic aspects of geometrically-necessary and statistically-stored dislocation density. *Acta Materialia, 47*, 1597–1611.

Asaro, R. J., & Needleman, A. (1985). Texture development and strain hardening in rate dependent polycrystals. *Acta Metallurgica, 33*, 923–953.

Ashby, M. F. (1970). The deformation of plastically non-homogeneous materials. *Philosophical Magazine, 21*, 399–424.

Barbe, F., Decker, L., Jeulin, D., & Cailletaud, G. (2001). Intergranular and intragranular behavior of polycrystalline aggregates. Part 1: F.E. model. *International Journal of Plasticity, 17*, 513–536.

Barbe, F., Forest, S., & Cailletaud, G. (2001). Intergranular and intragranular behavior of polycrystalline aggregates. Part 2: Results. *International Journal of Plasticity, 17*(4), 537–563.

Bardella, L. (2006). A deformation theory of strain gradient crystal plasticity that accounts for geometrically necessary dislocations. *Journal of the Mechanics and Physics of Solids, 54*, 128–160.

Bardella, L., Segurado, J., Panteghini, A., & Llorca, J. (2013). Latent hardening size effect in small-scale plasticity. *Modelling and Simulation in Materials Science and Engineering, 21*, 055009.

Bargieł, M., & Mościński, J. (1991). C-language program for the irregular close packing of hard spheres. *Computer Physics Communications, 64*, 183–192.

Barton, N. R., Knap, J., Arsenlis, A., Becker, R., Hornung, P. D., & Jefferson, D. R. (2008). Embedded polycrystal plasticity and adaptive sampling. *International Journal of Plasticity, 24*, 242–266.

Bassani, J. L., & Wu, T. Y. (1991). Latent hardening in single crystals II. Analytical characterization and predictions. *Proceedings of the Royal Society of London A*, *435*, 2141.

Bayerschen, E., McBride, A. T., Reddy, B. D., & Böhlke, T. (2016). Review on slip transmission criteria in experiments and crystal plasticity models. *Journal of Materials Science*, *51*, 2243–2258.

Bayley, C. J., Brekelmans, W. A. M., & Geers, M. G. D. (2006). A comparison of dislocation induced back stress formulations in strain gradient crystal plasticity. *International Journal of Solids and Structures*, *43*, 7268–7286.

Becker, R. (1991). Analysis of texture evolution in channel die compression—I. Effects of grain interaction. *Acta Metallurgica et Materialia*, *39*, 1211–1230.

Bennett, V. P., & McDowell, D. L. (2003). Polycrystal orientation distribution effects on microslip in high cycle fatigue. *International Journal of Fatigue*, *25*, 27–39.

Benzerga, A. A., & Leblond, J. B. (2010). Ductile fracture by void growth to coalescence. *Advances in Applied Mechanics*, *44*, 169–305.

Berbenni, S., Taupin, V., Djaka, K. S., & Fressengeas, C. (2014). A numerical spectral approach for solving elasto-static field dislocation and g-disclination mechanics. *International Journal of Solids and Structures*, *51*, 4157–4175.

Bertin, N., Tomé, C. N., Beyerlein, I. J., Barnett, M. R., & Capolungo, L. (2014). On the strength of dislocation interactions and their effect on latent hardening in pure magnesium. *International Journal of Plasticity*, *62*, 72–92.

Bertin, N., Upadhyay, M. V., Pradalier, C., & Capolungo, L. (2015). A FFT-based formulation for efficient mechanical fields computation in isotropic and anisotropic periodic discrete dislocation dynamics. *Modelling and Simulation in Materials Science and Engineering*, *23*(6), 065009.

Berveiller, M., & Zaoui, A. (1979). An extension to the self-consistent scheme to plastically-flowing polycristals. *Journal of the Mechanics and Physics of Solids*, *26*, 325–344.

Beyerlein, I. J., & Tomé, C. (2008). A dislocation-based constitutive law for pure Zr including temperature effects. *International Journal of Plasticity*, *24*, 867–895.

Bhattacharyya, A., El-Danaf, E., Kalidindi, S. R., & Doherty, R. D. (2001). Evolution of grain-scale microstructure during large strain simple compression of polycrystalline aluminum with quasi-columnar grains: OIM measurements and numerical simulations. *International Journal of Plasticity*, *17*, 861–883.

Bieler, T. R., Eisenlohr, P., Zhang, C., Phukan, H. J., & Crimp, M. A. (2014). Grain boundaries and interfaces in slip transfer. *Current Opinion in Solid State and Materials Science*, *18*, 212–226.

Bilby, B. A., & Crocker, A. G. (1965). The theory of the crystallography of deformation twinning. *Proceedings of the Royal Society of London A: Mathematical, Physical and Engineering Sciences*, *288*, 240–255.

Birosca, S., Buffiere, J.-Y., Garcia-Pastor, F., Karadge, M., Babout, L., & Preuss, M. (2009). Three-dimensional characterization of fatigue cracks in Ti-6246 using X-ray tomography and electron backscatter diffraction. *Acta Materialia*, *57*, 5834–5847.

Bobeth, M., & Diener, G. (1987). Static elastic and thermoelastic field fluctuations in multiphase composites. *Journal of the Mechanics and Physics of Solids*, *35*, 137–149.

Boehlert, C. J., Chen, Z., Gutiérrez-Urrutia, I., Llorca, J., & Pérez-Prado, M. T. (2012). In situ analysis of the tensile and tensile-creep deformation mechanisms in rolled AZ31. *Acta Materialia*, *60*, 1889–1904.

Böhm, H. J. (2004). *A short introduction to continuum micromechanics*. Springer-Verlag Wien.

Böhm, H. J., & Han, W. (2001). Comparisons between three-dimensional and two-dimensional multi-particle unit cell models for particle reinforced metal matrix composites. *Modelling and Simulation in Materials Science and Engineering*, *9*, 47–66.

Borg, U., Niordson, C. F., & Kysar, J. W. (2008). Size effects on void growth in single crystals with distributed voids. *International Journal of Plasticity*, *24*, 688–701.

Brenner, R., Beaudoin, A. J., Suquet, P., & Acharya, A. (2014). Numerical implementation of static Field Dislocation Mechanics theory for periodic media. *Philosophical Magazine*, *94*, 1764–1787.

Brenner, R., Lebensohn, R. A., & Castelnau, O. (2009). Elastic anisotropy and yield surface estimates of polycrystals. *International Journal of Solids and Structures*, *46*, 3018–3026.

Brisard, S., & Dormieux, L. (2010). FFT-based methods for the mechanics of composites: A general variational framework. *Computational Materials Science*, *49*, 663–671.

Bronkhorst, C. A., Kalidindi, S. R., & Anand, L. (1992). Polycrystalline plasticity and the evolution of crystallographic texture in FCC metals. *Philosophical Transactions: Physical Sciences and Engineering*, *341*, 443–477.

Busso, E. P., Meissonnier, F. T., & O'Dowd, N. P. (2000). Gradient-dependent deformation of two-phase single crystals. *Journal of the Mechanics and Physics of Solids*, *48*, 2333–2361.

Byer, C. M., & Ramesh, K. T. (2013). Effects of the initial dislocation density on size effects in single-crystal magnesium. *Acta Materialia*, *61*, 3808–3818.

Cailletaud, G. (1992). A micromechanical approach to inelastic behaviour of metals. *International Journal of Plasticity*, *8*, 55–73.

Capolungo, L., Beyerlein, I. J., & Tomé, C. N. (2009). Slip-assisted twin growth in hexagonal close-packed metals. *Scripta Materialia*, *60*, 32–35.

CAPSUL. (2018). https://materials.imdea.org/research/simulation-tools/capsul/.

Castelluccio, G. M., & McDowell, D. L. (2013). A mesoscale approach for growth of 3D microstructurally small fatigue cracks in polycrystals. *International Journal of Damage Mechanics*, *23*, 791–818.

Castelluccio, G. M., & McDowell, D. L. (2014). Mesoscale modeling of microstructurally small fatigue cracks in metallic polycrystals. *Materials Science and Engineering: A*, *598*, 34–55.

Castelluccio, G. M., & McDowell, D. L. (2015). Microstructure and mesh sensitivities of mesoscale surrogate driving force measures for transgranular fatigue cracks in polycrystals. *Materials Science and Engineering: A*, *639*, 626–639.

Cepeda-Jiménez, C. M., Molina-Aldareguia, J. M., & Pérez-Prado, M. T. (2015). Effect of grain size on slip activity in pure magnesium polycrystals. *Acta Materialia*, *84*, 443–456.

Cermelli, P., & Gurtin, M. E. (2001). On the characterization of geometrically necessary dislocations in finite plasticity. *Journal of the Mechanics and Physics of Solids*, *49*, 1539–1568.

Cerrone, A., Stein, C., Pokharel, R., Hefferan, C., Lind, J., Tucker, H., … Ingraffea, A. (2015). Implementation and verification of a microstructure-based capability for modeling microcrack nucleation in LSHR at room temperature. *Modelling and Simulation in Materials Science and Engineering*, *23*, 035006.

Chakraborty, A., & Eisenlohr, P. (2017). Evaluation of an inverse methodology for estimating constitutive parameters in face-centered cubic materials from single crystal indentations. *European Journal of Mechanics - A/Solids*, *66*, 114–124.

Chang, Y., & Kochmann, D. M. (2015). A variational constitutive model for slip-twinning interactions in hcp metals: Application to single- and polycrystalline magnesium. *International Journal of Plasticity*, *73*, 39–61.

Charkaluk, E., Bignonnet, A., Constantinescu, A., & Dang Van, K. (2002). Fatigue design of structures under thermomechanical loadings. *Fatigue and Fracture of Engineering Materials and Structures*, *25*, 1199–1206.

Chen, B., Jiang, J., & Dunne, F. P. (2018). Is stored energy density the primary meso-scale mechanistic driver for fatigue crack nucleation? *International Journal of Plasticity*, *101*, 213–229.

Cheng, J., & Ghosh, S. (2017). Crystal plasticity finite element modeling of discrete twin evolution in polycrystalline magnesium. *Journal of the Mechanics and Physics of Solids*, *99*, 512–538.

Cheng, J., Shen, J., Mishra, R. K., & Ghosh, S. (2018). Discrete twin evolution in Mg alloys using a novel crystal plasticity finite element model. *Acta Materialia, 149*, 142–153.
Cheong, K. S., & Busso, E. P. (2004). Discrete dislocation density modelling of single phase FCC polycrystal aggregates. *Acta Materialia, 52*, 5665–5675.
Cheong, K. S., Busso, E. P., & Arsenlis, A. (2005). A study of microstructural length scale effects on the behaviour of FCC polycrystals using strain gradient concepts. *International Journal of Plasticity, 21*, 1797–1814.
Christian, J. W., & Mahajan, S. (1995). Deformation twinning. *Progress in Materials Science, 39*, 1–157.
Chuan, W., He, Y., & Wei, L. H. (2013). Modeling of discontinuous dynamic recrystallization of a near-β titanium alloy IMI834 during isothermal hot compression by combining a cellular automaton model with a crystal plasticity finite element method. *Computational Materials Science, 79*, 944–959.
Cruzado, A., Gan, B., Jimnez, M., Barba, D., Ostolaza, K., Linaza, A., … Segurado, J. (2015). Multiscale modeling of the mechanical behavior of IN718 superalloy based on micropillar compression and computational homogenization. *Acta Materialia, 98*, 242–253.
Cruzado, A., LLorca, J., & Segurado, J. (2017). Modeling cyclic deformation of inconel 718 superalloy by means of crystal plasticity and computational homogenization. *International Journal of Solids and Structures, 122–123*, 148–161.
Cruzado, A., Lucarini, S., LLorca, J., & Segurado, J. (2018a). Crystal-plasticity simulation of the effect grain size on the fatigue behavior of polycrystalline Inconel 718. *International Journal of Fatigue, 113*, 236–245.
Cruzado, A., Lucarini, S., LLorca, J., & Segurado, J. (2018b). Microstructure-based fatigue life model of metallic alloys with bilinear Coffin-Manson behavior. *International Journal of Fatigue, 107*, 40–48.
Cuitiño, A., & Ortiz, M. (1992). Material-independent method for extending stress update algorithms from small-strain plasticity to finite plasticity with multiplicative kinematics. *Engineering Computations, 9*, 437–451.
Dai, H., & Parks, D. M. (1997). Geometrically-necessary dislocation density and scale-dependent crystal plasticity. In A. Khan (Ed.), *Proceedings of sixth international symposium on plasticity* (pp. 17–18): Gordon & Breach [Ed.].
Damask. (2018). https://damask.mpie.de.
de Botton, G., & Ponte Castañeda, P. (1995). Variational estimates for the creep behavior of polycrystals. *Proceedings of the Royal Society of London A, 448*, 121–142.
de Geus, T. W. J., Vondrejc, J., Zeman, J., Peerlings, R. H. J., & Geers, M. G. D. (2017). Finite strain FFT-based non-linear solvers made simple. *Computer Methods in Applied Mechanics and Engineering, 318*, 412–430.
Delaire, F., Raphanel, J. L., & Rey, C. (2000). Plastic heterogeneities of a copper multicrystal deformed in uniaxial tension: Experimental study and finite element simulations. *Acta Materialia, 48*, 1075–1087.
de Sansal, C., Devincre, B., & Kubin, L. P. (2010). Grain size strengthening in microcrystalline copper: A three-dimensional dislocation dynamics simulation. *Key Engineering Materials, 423*, 25–32.
de Souza Neto, E. A., Peric, D., & Owen, D. R. J. (2008). *Computational methods for plasticity*. John Wiley and Sons, Ltd.
Devincre, B., Hoc, T., & Kubin, L. (2008). Dislocation mean free paths and strain hardening of crystals. *Science, 320*, 1745–1748.
Doškář, M., Novák, J., & Zeman, J. (2014). A periodic compression and reconstruction of real-world material systems based on Wang tiles. *Physical Review E, 90*, 062118.
Dream.3D. (2018). http://dream3d.bluequartz.net.
Dunne, F. P. E., Rugg, D., & Walker, A. (2007). Length scale-dependent, elastically anisotropic, physically-based hcp crystal plasticity: Application to cold-dwell fatigue in Ti alloys. *International Journal of Plasticity, 23*, 1061–1083.

Echlin, M. P., Husseini, N. S., Nees, J. A., & Pollock, T. M. (2011). A new femtosecond laser-based tomography technique for multiphase materials. *Advanced Materials*, *23*, 2339–2342.

Eisenlohr, P., Diehl, M., Lebensohn, R. A., & Roters, F. (2013). A spectral method solution to crystal elasto-viscoplasticity at finite strains. *International Journal of Plasticity*, *46*, 37–53.

El-Awady, J. A. (2015). Unraveling the physics of size-dependent dislocation-mediated plasticity. *Nature Communications*, *6*, 5926.

Escobedo, J. P., Cerreta, E. K., Dennis-Koller, D., Patterson, B. M., Bronkhorst, C., Hansen, B., ... Lebensohn, R. A. (2011). Effects of grain size and boundary structure on the dynamic tensile response of copper. *Journal of Applied Physics*, *110*, 033513.

Escobedo, J. P., Cerreta, E. K., Dennis-Koller, D., Trujillo, C. P., & Bronkhorst, C. (2013). Influence of boundary structure and near neighbor crystallographic orientation on the dynamic damage evolution during shock loading. *Philosophical Magazine*, *93*, 833–846.

Eshelby, J. D. (1957). The determination of the elastic field of an ellipsoidal inclusion and related problems. *Proceedings of the Royal Society of London A*, *252*, 561–569.

Estrin, Y., & Mecking, H. (1984). A unified phenomenological description of work hardening and creep based on one-parameter models. *Acta Metallurgica*, *32*, 57–70.

Eswar Prasad, K., Rajesh, K., & Ramamurty, U. (2014). Micropillar and macropillar compression responses of magnesium single crystals oriented for single slip or extension twinning. *Acta Materialia*, *65*, 316–325.

Evers, L. P., Brekelmans, W. A. M., & Geers, M. G. D. (2004). Non-local crystal plasticity model with intrinsic SSD and GND effects. *Journal of the Mechanics and Physics of Solids*, *52*, 2379–2401.

Eyre, D. J., & Milton, G. W. (1999). A fast numerical scheme for computing the response of composites using grid refinement. *The European Physical Journal Applied Physics*, *6*, 41–47.

Fatemi, A., & Socie, D. F. (1988). A critical plane approach to multiaxial fatigue damage including out-of-phase loading. *Fatigue and Fracture of Engineering Materials and Structures*, *11*, 149–165.

Fatemi, A., & Socie, D. F. (1989). Multiaxial fatigue: Damage mechanisms and life predictions. *Advances in fatigue science and technology. NATO ASI series (Series E: Applied Sciences): Vol. 159* (pp. 877–890): Springer.

Fayman, Y. C. (1987). Microstructural characterization and element partitioning in a direct-aged superalloy (DA718). *Materials Science and Engineering A*, *92*, 159–171.

Fernández, A., Jérusalem, A., Gutiérrez-Urrutia, I., & Pérez-Prado, M. (2013). Three-dimensional investigation of the grain boundary-twin interactions in a Mg AZ31 alloy by electron backscatter diffraction and continuum modeling. *Acta Materialia*, *61*, 7679–7692.

Feyel, F. (2003). A multilevel finite element method (FE2) to describe the response of highly non-linear structures using generalized continua. *Compuational Methods in Applied Mechanics and Engineering*, *192*, 3233–3244.

Fleck, N. A., Muller, G. M., Ashby, M. F., & Hutchinson, J. W. (1994). Strain gradient plasticity: Theory and experiment. *Acta Metallurgica et Materialia*, *42*, 475–487.

Franciosi, P., & Zaoui, A. (1982). Multislip in f.c.c. crystals a theoretical approach compared with experimental data. *Acta Metallurgica*, *30*, 1627–1637.

Fullwood, D. T., Niezgoda, S. R., & Kalidindi, S. R. (2008). Microstructure reconstructions from 2-point statistics using phase-recovery algorithms. *Acta Materialia*, *56*, 942–948.

Geers, M., Cottura, M., Appolaire, B., Busso, E. P., Forest, S., & Villani, A. (2014). Coupled glide-climb diffusion-enhanced crystal plasticity. *Journal of the Mechanics and Physics of Solids*, *70*, 136–153.

Ghosh, S., & Dimiduk, D. (2011). *Computational methods for microstructure property relationship*. Springer.

Ghosh, S., Shahba, A., Tu, X., Huskins, E. L., & Schuster, B. E. (2016). Crystal plasticity FE modeling of Ti alloys for a range of strain-rates. Part II: Image-based model with experimental validation. *International Journal of Plasticity*, *87*, 69–85.

Gong, J., & Wilkinson, A. J. (2011). A microcantilever investigation of size effect, solid-solution strengthening and second-phase strengthening for a prism slip in alpha-Ti. *Acta Materialia*, *59*, 5970–5981. https://doi.org/10.1016/j.actamat.2011.06.005.

González, C., & LLorca, J. (2007). Mechanical behavior of unidirectional fiber-reinforced polymers under transverse compression: Microscopic mechanisms and modeling. *Composites Science and Technology*, *67*, 2795–2806.

González, C., Segurado, J., & LLorca, J. (2004). Numerical simulation of elasto-plastic deformation of composites: Evolution of stress microfields and implications for homogenization models. *Journal of the Mechanics and Physics of Solids*, *52*, 1573–1593.

González, C., Vilatela, J. J., Molina-Aldareguía, J. M., Lopes, C. S., & LLorca, J. (2017). Structural composites for multifunctional applications: Current challenges and future trends. *Progress in Materials Science*, *89*, 194–251.

Graham, J. T., Rollett, A. D., & Lesar, R. (2016). Fast fourier transform discrete dislocation dynamics. *Modelling and Simulation in Materials Science and Engineering*, *24*, 085005.

Grennerat, F., Montagnat, M., Castelnau, O., Vacher, P., Moulinec, H., Suquet, P., & Duval, P. (2012). Experimental characterization of the intragranular strain field in columnar ice during transient creep. *Acta Materialia*, *60*, 3655–3666.

Gurtin, M. E. (2002). A gradient theory of single-crystal viscoplasticity that accounts for geometrically necessary dislocations. *Journal of the Mechanics and Physics of Solids*, *50*, 5–32.

Gurtin, M. E. (2008). A finite-deformation, gradient theory of single-crystal plasticity with free energy dependent on densities of geometrically necessary dislocations. *International Journal of Plasticity*, *24*, 702–725.

Gurtin, M. E., & Needleman, A. (2005). Boundary conditions in small-deformation, single-crystal plasticity that account for the burgers vector. *Journal of the Mechanics and Physics of Solids*, *53*, 1–31.

Gvenc, O., Bambach, M., & Hirt, G. (2014). Coupling of crystal plasticity finite element and phase field methods for the prediction of SRX kinetics after hot working. *Steel Research International*, *85*, 999–1009.

Hall, E. O. (1951). The deformation and ageing of mild steel: III Discussion of results. *Proceedings of the Physical Society Section B*, *64*, 747–753.

Han, C., Ma, A., Roters, F., & Raabe, D. (2007). A finite element approach with patch projection for strain gradient plasticity formulations. *International Journal of Plasticity*, *23*, 690–710.

Han, C.-S., Gao, H., Huang, Y., & Nix, W. D. (2005a). Mechanism-based strain gradient crystal plasticity I. Theory. *Journal of the Mechanics and Physics of Solids*, *53*, 1188–1203.

Han, C.-S., Gao, H., Huang, Y., & Nix, W. D. (2005b). Mechanism-based strain gradient crystal plasticity II. Analysis. *Journal of the Mechanics and Physics of Solids*, *53*, 1204–1222.

Hashin, Z., & Shtrikman, S. (1963). A variational approach to the theory of elastic behavior of multiphase materials. *Journal of the Mechanics and Physics of Solids*, *11*, 127–140.

Hasija, V., Ghosh, S., Mills, M. J., & Joseph, D. S. (2003). Deformation and creep modeling in polycrystalline Ti–6Al alloys. *Acta Materialia*, *51*, 4533–4549.

Hauoala, S., Segurado, J., & LLorca, J. (2018). An analysis of the influence of grain size on the strength of FCC polycrystals by means of computational homogenization. *Acta Materialia*, *148*, 72–85.

Hazanov, S., & Huet, C. (1994). Order relationships for boundary condition effects in heterogeneous bodies smaller that the representative volume. *Journal of the Mechanics and Physics of Solids*, *42*, 1995–2011.

Heilbronner, R., & Bruhn, D. (1998). The influence of three-dimensional grain size distributions on the rheology of polyphase rocks. *Journal of Structural Geology*, *20*, 695–707.

Heinz, A., & Neumann, P. (1990). Crack initiation during high cycle fatigue of an austenitic steel. *Acta Metallurgica et Materialia*, *38*, 1933–1940.

Hémery, S., Nizou, P., & Villechaise, P. (2018). In situ SEM investigation of slip transfer in Ti-6Al-4V: Effect of applied stress. *Materials Science and Engineering: A*, *709*, 277–284.

Herbig, M., King, A., Reischig, P., Proudhon, H., Lauridsen, E. M., Marrow, J., … Ludwig, W. (2011). 3-D growth of a short fatigue crack within a polycrystalline microstructure studied using combined diffraction and phase-contrast X-ray tomography. *Acta Materialia*, *59*, 590–601.

Herrera-Solaz, V., Hidalgo-Manrique, P., Pérez-Prado, M. T., Letzig, D., LLorca, J., & Segurado, J. (2014). Effect of rare earth additions on the critical resolved shear stresses of magnesium alloys. *Materials Letters*, *128*, 199–203.

Herrera-Solaz, V., LLorca, J., Dogan, E., Karaman, I., & Segurado, J. (2014). An inverse optimization strategy to determine single crystal mechanical behavior from polycrystal tests: Application to AZ31 Mg alloy. *International Journal of Plasticity*, *57*, 1–15.

Herrera-Solaz, V., Segurado, J., & LLorca, J. (2015). On the robustness of an inverse optimization approach based on the Levenberg-Marquardt method for the mechanical behavior of polycrystals. *European Journal of Mechanics/A*, *53*, 220–228.

Hershley, A. V. (1954). The elasticity of an isotropic aggregate of anisotropic cubic crystal. *Journal of Applied Mechanics*, *21*, 236–240.

Hidalgo-Manrique, P., Herrera-Solaz, V., Segurado, J., LLorca, J., Gálvez, F., Ruano, O. A., … Pérez-Prado, M. T. (2015). Origin of the reversed yield asymmetry in Mg-rare earth alloys at high temperature. *Acta Materialia*, *92*, 265–277.

Hill, R. (1965a). Continuum micro-mechanics of elastoplastic polycrystals. *Journal of the Mechanics and Physics of Solids*, *13*, 89–101.

Hill, R. (1965b). A self-consistent mechanics of composite materials. *Journal of the Mechanics and Physics of Solids*, *13*, 213–222.

Hill, R. (1966). Generalized constitutive relations for incremental deformation of metal crystals by multislip. *Journal of the Mechanics and Physics of Solids*, *14*, 95–102.

Hill, R., & Rice, J. R. (1972). Constitutive analysis of elastic-plastic crystals at arbitrary strain. *Journal of the Mechanics and Physics of Solids*, *20*, 401–413.

Hirth, J. P., & Lothe, J. (1982). *Theory of dislocations*. Krieger Publishing Company.

Hong, X., Godfrey, A., & Liu, W. (2016). Challenges in the prediction of twin transmission at grain boundaries in a magnesium alloy. *Scripta Materialia*, *123*, 77–80.

Houtte, P. V. (1978). Simulation of the rolling and shear texture of brass by the taylor theory adapted for mechanical twinning. *Acta Metallurgica*, *26*, 591–604. https://doi.org/10.1016/0001-6160(78)90111-6.

Hu, Z., Rauch, E. F., & Teodosiu, C. (1992). Work-hardening behavior of mild steel under stress reversal at large strains. *International Journal of Plasticity*, *8*, 839–856.

Hutchinson, J. (1976). Bounds and self-consistent estimates for creep of polycrystalline materials. *Proceedings of the Royal Society of London A*, *348*, 101–127.

Idiart, M. I., Moulinec, H., & Ponte Castañeda, P. (2006). Macroscopic behavior and field fluctuations in viscoplastic composites: Second-order estimates versus full-field simulations. *Journal of the Mechanics and Physics of Solids*, *54*, 1029–1063.

Jeulin, D., Kanit, T., & Forest, S. (2004). Representative volume element: A statistical point of view. In D. J. Bergman & E. Inan (Eds.), *Continuum models and discrete systems* (pp. 21–27): Springer [Eds.].

Jung, J., Yoon, J. I., Kim, J., Latypov, M., Kim, J., & Kim, H. S. (2017). Continuum understanding of twin formation near grain boundaries of FCC metals with low stacking fault energy. *npj Computational Materials*, *3*, 21.

Joseph, D. S., Chakraborty, P., & Ghosh, S. (2010). Wavelet transformation based multi-time scaling method for crystal plasticity FE simulations under cyclic loading. *Computer Methods in Applied Mechanics and Engineering*, *199*, 2177–2194.

Kabel, M., Böhlke, T., & Schneider, M. (2014). Efficient fixed point and Newton–Krylov solvers for FFT-based homogenization of elasticity at large deformations. *Computational Mechanics*, *54*, 1497–1514.

Kalidindi, S. R. (1998). Incorporation of deformation twinning in crystal plasticity models. *Journal of the Mechanics and Physics of Solids*, *46*, 267–271 and 273–290.

Kalidindi, S. R., Bronkhorst, C. A., & Anand, L. (1992). Crystallographic texture evolution in bulk deformation processing of FCC metals. *Journal of the Mechanics and Physics of Solids*, *40*, 537–569.

Kalidindi, S. R., Knezavic, M., Levinson, A., Harris, R., Mishra, R. J., & Doherty, R. D. (2010). Deformation twinning in AZ31: Influence on strain hardening and texture evolution. *Acta Materialia*, *58*, 6230–6242.

Khan, R., Zahedi, F. I., & Siddiqui, A. K. (2016). Numerical modeling of twinning induced plasticity in austenite based advanced high strength steels. *Procedia Manufacturing*, *5*, 772–786.

Kiener, D., Grosinger, W., Dehm, G., & Pippan, R. (2008). A further step towards an understanding of size-dependent crystal plasticity: In situ tension experiments of miniaturized single-crystal copper samples. *Acta Materialia*, *56*, 580–592.

Knezevic, M., & Beyerlein, I. J. (2018). Multiscale modeling of microstructure-property relationships of polycrystalline metals during thermo-mechanical deformation. *Advanced Engineering Materials*, *20*, 1700956.

Knezevic, M., Crapps, J., Beyerlein, I. J., Coughlin, D. R., Clarke, K. D., & McCabe, R. J. (2016). Anisotropic modeling of structural components using embedded crystal plasticity constructive laws within finite elements. *International Journal of Mechanical Sciences*, *105*, 227–238.

Knezevic, M., Zecevic, M., Beyerlein, I. J., & Lebensohn, R. A. (2016). A numerical procedure enabling accurate descriptions of strain rate-sensitive flow of polycrystals within crystal visco-plasticity theory. *Computational Methods in Applied Mechanics and Engineering*, *308*, 468–482.

Kocks, W. (1975). Thermodynamics and kinetics of slip. *Progress in Materials Science*, *19*, 1–291.

Konrad, J., Zaefferer, S., & Raabe, D. (2006). Investigation of orientation gradients around a hard laves particle in a warm-rolled Fe_3Al-based alloy using a 3D EBSD-FIB technique. *Acta Materialia*, *54*, 1369–1380.

Korsunsky, A. M., Dini, D., Dunne, F. P. E., & Walsh, M. J. (2007). Comparative assessment of dissipated energy and other fatigue criteria. *International Journal of Fatigue*, *29*, 1990–1995.

Kothari, M., & Anand, L. (1998). Elasto-viscoplastic constitutive equations for polycrystalline metals: Application to tantalum. *Journal of the Mechanics and Physics of Solids*, *46*, 51–83.

Kothary, M., & Anand, L. (1998). Elasto-viscoplastic constitutive equations for polycrystalline metals: Application to tantalum. *Journal of the Mechanics and Physics of Solids*, *46*, 51–67. 69–83.

Kouznetsova, V. G., & Geers, M. G. D. (2008). A multi-scale model of martensitic transformation plasticity. *Mechanics of Materials*, *40*, 641–657.

Kowalczyk-Gajewska, K. (2010). Modelling of texture evolution in metals accounting for lattice reorientation due to twinning. *European Journal of Mechanics - A/Solids*, *29*, 28–41.

Kreher, W. (1990). Residual-stresses and stored elastic energy of composites and polycrystals. *Journal of the Mechanics and Physics of Solids*, *38*, 115–128.

Kröner, E. (1958). Berechnung der Elastischen Konstanten des Vielkristalls aus den Konstanten des Einkristalls. *Zeitung Physik*, *51*, 504–518.

Kumar, M. A., Beyerlein, I. J., McCabe, R. J., & Tomé, C. N. (2016). Grain neighbour effects on twin transmission in hexagonal close packed materials. *Nature Communications*, 7, 13826.

Lahellec, N., Michel, J. C., Moulinec, H., & Suquet, P. (2001). Analysis of inhomogeneous materials at large strains using fast Fourier transforms. In C. Miehe (Ed.), *Iutam symposium on computational mechanics of solids materials* (pp. 247–268): Kluwer Academic [Ed.].

Lautensack, C. (2007). *Random Laguerre tessellations* (PhD thesis). Universitt Karlsruhe.

Laws, N. (1973). On the thermostatics of composite materials. *Journal of the Mechanics and Physics of Solids, 21*, 9–17.

Lebensohn, R. A. (2001). N-site modelling of a 3D viscoplastic polycrystal using fast Fourier transform. *Acta Materialia, 49*, 2723–2737.

Lebensohn, R. A., Brenner, R., Castelnau, O., & Rollett, A. D. (2008). Orientation image-based micromechanical modelling of subgrain texture evolution in polycrystalline copper. *Acta Materialia, 56*, 3914–3926.

Lebensohn, R. A., & Canova, G. R. (1997). Selfconsistent approach for modelling texture development of two-phase polycrystals: Application to Titanium alloys. *Acta Materialia, 45*, 3687–3694.

Lebensohn, R. A., Castelnau, O., Brenner, R., & Gilormini, P. (2005). Study of the antiplane deformation of linear 2-D polycrystals with different microstructures. *International Journal of Solids and Structures, 42*, 5441–5449.

Lebensohn, R. A., Escobedo, J. P., Cerreta, E. K., Dennis-Koller, D., Bronkhorst, C. A., & Bingert, J. (2013). Modelling void growth in polycrystalline materials. *Acta Materialia, 61*, 6918–6932.

Lebensohn, R. A., Hartley, C. S., Tomé, C., & Castelnau, O. (2010). Modelling the mechanical response of polycrystals deforming by climb and glide. *Philosophical Magazine, 90*, 567–583.

Lebensohn, R. A., Idiart, M. I., Ponte Castañeda, P., & Vincent, P. G. (2011). Dilatational viscoplasticity of polycrystalline solids with intergranular cavities. *Philosophical Magazine, 91*, 3038–3067.

Lebensohn, R. A., Kanjarla, A. K., & Eisenlohr, P. (2012). An elasto-viscoplastic formulation based on fast fourier transforms for the prediction of micromechanical fields in polycrystalline materials. *International Journal of Plasticity, 32–33*, 59–69.

Lebensohn, R. A., Montagnat, M., Mansuy, P., Duval, P., Meysonnier, J., & Philip, A. (2009). Modeling viscoplastic behavior and heterogenous intracrystalline deformation of columnar ice polycrystals. *Acta Materialia, 57*, 1405–1415.

Lebensohn, R. A., & Needleman, A. (2016). Numerical implementation of non-local polycrystal plasticity using fast Fourier transforms. *Journal of the Mechanics and Physics of Solids, 97*, 333–351.

Lebensohn, R. A., & Tomé, C. N. (1993). A self-consistent anisotropic approach for the simulation of plastic deformation and texture development of polycrystals: Application to zirconium alloys. *Acta Metallurgica et Materialia, 41*, 2611 2624.

Lebensohn, R. A., Tomé, C. N., & Ponte Castañeda, P. (2007). Self-consistent modeling of the mechanical behavior of viscoplastic polycrystals incorporating intragranular field fluctuations. *Philosophical Magazine, 87*, 4287–4322.

Lebensohn, R. A., Uhlenhut, H., Hartig, C., & Mecking, H. (1998). Mechanical behavior gamma-TiAl-based polysinthetically twinned crystals: Micromechanical modelling and experimental validation. *Acta Materialia, 46*, 4701–4709.

Lebensohn, R. A., Zecevic, M., Knezevic, M., & McCabe, R. J. (2016). Average intragranular misorientation trends in polycrystalline materials predicted by a viscoplastic self-consistent approach. *Acta Materialia, 104*, 228–236.

Lecarme, L., Maire, E., Kumar, K. C. A., De Vleeschouwer, C., Jacques, L., Simar, A., & Pardoen, T. (2014). Heterogenous void growth revealed by in situ 3-D X-ray microtomography using automatic cavity tracking. *Acta Materialia, 63*, 130–139.

Lee, E. H., & Liu, D. T. (1967). Finite-strain elastic–plastic theory with application to plane-wave analysis. *Journal of Applied Physics, 38*, 19–27.

Lele, S. P., & Anand, L. (2008). A small-deformation strain-gradient theory for isotropic viscoplastic materials. *Philosophical Magazine, 88*, 3655–3689.

Li, H., Sun, X., & Yang, H. (2016). A three-dimensional cellular automata-crystal plasticity finite element model for predicting the multiscale interaction among heterogeneous deformation, DRX microstructural evolution and mechanical responses in titanium alloys. *International Journal of Plasticity, 87*, 154–180.

Lim, H., Abdeljawad, F., Owen, S. J., Hanks, B. W., Foulk, J. W., & Battaile, C. C. (2016). Incorporating physically-based microstructures in materials modeling: Bridging phase field and crystal plasticity frameworks. *Modelling and Simulation in Materials Science and Engineering, 24*, 045016.

Liu, C., Shanthraj, P., Diehl, M., Roters, F., Dong, S., Dong, J., … Raabe, D. (2018). An integrated crystal plasticity–phase field model for spatially resolved twin nucleation, propagation, and growth in hexagonal materials. *International Journal of Plasticity, 106*, 203–227.

Liu, Y., Li, N., Arul Kumar, M., Pathak, S., Wang, J., McCabe, R. J., … Tomé, C. N. (2017). Experimentally quantifying critical stresses associated with basal slip and twinning in magnesium using micropillars. *Acta Materialia, 135*, 411–421.

Liu, Y., & Ponte Castañeda, P. (2004). Second-order theory for the effective behavior and field fluctuations of polycrystals. *Journal of the Mechanics and Physics of Solids, 52*, 467–495.

LLorca, J., González, C., Molina-Aldareguía, J. M., Segurado, J., Seltzer, R., Sket, F., … Canal, L. P. (2011). Multiscale modeling of composite materials: A roadmap towards virtual testing. *Advanced Materials, 23*, 5130–5147.

LLorca, J., & Segurado, J. (2004). Three-dimensional multiparticle cell simulations of deformation and damage in sphere-reinforced composites. *Materials Science and Engineering A, 365*, 267–274.

Lucarini, S., & Segurado, J. (2018). On the accuracy of spectral solvers for micromechanics based fatigue modeling. *Computational Mechanics*, 1–18. https://doi.org/10.1007/s00466-018-1598-1.

Ma, A., & Roters, F. (2004). A constitutive model for fcc single crystals based on dislocation densities and its application to uniaxial compression of aluminium single crystals. *Acta Materialia, 52*, 3603–3612.

Ma, A., Roters, F., & Raabe, D. (2006). A dislocation density based constitutive model for crystal plasticity FEM including geometrically necessary dislocations. *Acta Materialia, 54*, 2169–2179.

Mahajan, S., & Williams, D. F. (1973). Deformation twinning in metals and alloys. *International Metallurgical Reviews, 18*, 43–61.

Malyar, N. V., Micha, J. S., Dehm, G., & Kirchlechner, C. (2017). Size effect in bi-crystalline micropillars with a penetrable high angle grain boundary. *Acta Materialia, 129*, 312–320.

Mandal, S., Lao, J., Donegan, S., & Rollett, A. D. (2018). Generation of statistically representative synthetic three-dimensional microstructures. *Scripta Materialia, 146*, 128–132.

Manonukul, A., & Dunne, F. P. E. (2004). High- and low-cycle fatigue crack initiation using polycrystal plasticity. *Proceedings of the Royal Society of London A: Mathematical, Physical and Engineering Sciences, 460*, 1881–1903.

Mareau, C., & Daymond, M. R. (2016). Micromechanical modelling of twinning in polycrystalline materials: Application to magnesium. *International Journal of Plasticity, 85*, 156–171.

Masson, R., Bornert, M., Suquet, P., & Zaoui, A. (2000). Affine formulation for the prediction of the effective properties of nonlinear composites and polycrystals. *Journal of the Mechanics and Physics of Solids, 48*, 1203–1227.

Mayeur, J. R., Beyerlein, I. J., Bronkhorst, C. A., & Mourad, H. M. (2015). Incorporating interface affected zones into crystal plasticity. *International Journal of Plasticity, 65*, 206–225.

McColl, I. R., Ding, J., & Leen, S. B. (2004). Finite element simulation and experimental validation of fretting wear. *Wear, 256*, 1114–1127.
McDowell, D. L., & Dunne, F. P. E. (2010). Microstructure-sensitive computational modeling of fatigue crack formation. *International Journal of Fatigue, 32*, 1521–1542.
McDowell, D. L., Gall, K., Horstemeyer, M. F., & Fan, J. (2003). Microstructure-based fatigue modeling of cast A356-T6 alloy. *Engineering Fracture Mechanics, 70*, 49–80.
Mecking, H., & Kocks, U. F. (1981). Kinetics of flow and strain-hardening. *Acta Metallurgica, 29*, 1865–1875.
Méric, L., Poubanne, P., & Cailletaud, G. (1991). Single crystal modeling for structural calculations: Part 1—Model presentation. *ASME Journal of Engineering Materials and Technology, 113*, 162–170.
Michel, J. C., Moulinec, H., & Suquet, P. (1999). Effective properties of composite materials with periodic microstructure: A computational approach. *Compuational Methods in Applied Mechanics and Engineering, 172*, 109–143.
Michel, J. C., Moulinec, H., & Suquet, P. (2000). A computational method based on augmented Lagrangians and fast Fourier transforms for composites with high contrast. *Computer Modeling in Engineering and Sciences, 1*, 79–88.
Michel, J. C., & Suquet, P. (2016). A model-reduction approach to the micromechanical analysis of polycrystalline materials. *Computational Mechanics, 57*, 483–508.
Miehe, C. (1996). Exponential map algorithm for stress updates in anisotropic multiplicative elastoplasticity for single crystals. *International Journal for Numerical Methods in Engineering, 39*, 3367–3390.
Miehe, C., Schotte, J., & Lambrecht, M. (2002). Homogenization of inelastic solid materials at finite strains based on incremental minimization principles. Application to the texture analysis of polycrystals. *Journal of the Mechanics and Physics of Solids, 50*, 2123–2167.
Miehe, C., Schrodër, J., & Schotte, J. (1999). Computational homogenization analysis in finite plasticity simulation of texture development in polycrystalline materials. *Computational Methods in Applied Mechanics and Engineering, 171*, 387–418.
Mika, D. P., & Dawson, P. R. (1999). Polycrystal plasticity modeling of intracrystalline boundary textures. *Acta Materialia, 47*, 1355–1369.
Molinari, A., Canova, G. R., & Ahzi, S. (1987a). Self-consistent approach of the large deformation polycrystal viscoplasticity. *Acta Metallurgica, 35*, 2983–2994.
Molinari, A., Canova, G. R., & Ahzi, S. (1987b). A self-consistent approach of the large deformation polycrystal viscoplasticity. *Acta Metallurgica, 35*, 2983–2994.
Monchiet, V., & Bonnet, G. (2013). Numerical homogenization of nonlinear composites with a polarization-based FFT iterative scheme. *Computational Materials Science, 79*, 276–283.
Mori, T., & Tanaka, K. (1973). Average stress in the matrix and average elastic energy of materials with misfitting inclusions. *Acta Metallurgica et Materialia, 21*, 571–574.
Motz, C., Schoeber, T., & Pippan, R. (2005). Mechanical properties of micro-sized copper bending beams machined by the focused ion beam technique. *Acta Materialia, 53*, 4269–4279.
Moulinec, H., & Silva, F. (2014). Comparison of three accelerated FFT-based schemes for computing the mechanical response of composite materials. *International Journal of Numerical Methods in Engineering, 97*, 960–985.
Moulinec, H., & Suquet, P. (1994). A fast numerical method for computing the linear and nonlinear mechanical properties of composites. *Comptes Rendus de l'Académie des Sciences, 318*, 1417–1423.
Moulinec, H., & Suquet, P. (1998). A numerical method for computing the overall response of nonlinear composites with complex microstructure. *Compuational Methods in Applied Mechanics and Engineering, 157*, 69–94.

Mueller, T., Kusne, A. G., & Ramprasad, R. (2017). Machine learning in materials science: Recent progress and emerging applications. *Reviews in Computational Chemistry, 29*. 1866–273.

Mura, T. (1987). *Micromechanics of defects in solids*. Martinus-Nijhoff Publishers.

Musinski, W. D., & McDowell, D. L. (2012). Microstructure-sensitive probabilistic modeling of HCF crack initiation and early crack growth in Ni-base superalloy IN100 notched components. *International Journal of Fatigue, 37*, 41–53.

Musinski, W. D., & McDowell, D. L. (2016). Simulating the effect of grain boundaries on microstructurally small fatigue crack growth from a focused ion beam notch through a three-dimensional array of grains. *Acta Materialia, 112*, 20–39.

Nabarro, F. R. N. (1947). Dislocations in a simple cubic lattice. *Proceedings of the Physical Society, 59*, 256.

Nayyeri, G., Poole, W. J., Sinclair, C. W., & Zaefferer, S. (2017). The role of indenter radius on spherical indentation of high purity magnesium loaded nearly parallel to the c-axis. *Scripta Materialia, 137*, 119–122.

Nemat-Nasser, S., & Hori, M. (1993). *Micromechanics: Overall properties of heterogeneous materials*. Elsevier Science Publishers.

Nemcko, M. J., & Wilkinson, D. S. (2016). On the damage and fracture of commercially pure magnesium using x-ray microtomography. *Materials Science and Engineering: A, 676*, 146–155.

Neper. (2018). http://neper.sourceforge.net.

Nguyen, V. D., Béchet, E., Geuzaine, C., & Noels, L. (2012). Imposing periodic boundary condition on arbitrary meshes by polynomial interpolation. *Computational Materials Science, 55*, 390–406.

Niezgoda, S. R., Turner, D. M., Fullwood, D. T., & Kalidindi, S. R. (2010). Optimized structure based representative volume element sets reflecting the ensemble-averaged 2-point statistics. *Acta Materialia, 58*, 4432–4445.

Niordson, C. F., & Kysar, J. W. (2014). Computational strain gradient crystal plasticity. *Journal of the Mechanics and Physics of Solids, 62*, 31–47.

Nix, W. D., & Gao, H. (1998). Indentation size effects in crystalline materials: A law for strain gradient plasticity. *Journal of the Mechanics and Physics of Solids, 46*, 411–425.

Nye, J. F. (1953). Some geometrical relations in dislocated crystals. *Acta Metallurgica, 1*, 153–162.

Ohno, N., & Wang, J.-D. (1993). Kinematic hardening rules with critical state of dynamic recovery, Part II: Application to experiments of ratcheting behavior. *International Journal of Plasticity, 9*, 391–403.

Olson, G. B. (2013). Genomic materials design: The ferrous frontier. *Acta Materialia, 61*, 771–781.

Orowan, E. (1934). Zur Kristallplastizität. *Zeitschrift für Physik, 89*, 605–613.

Ostien, J., & Garikipati, K. (2008). Discontinuous Galerkin method for an incompatibility-based strain gradient plasticity theory. In B. D. Reddy (Ed.), *Iutam symposium on theoretical, computational and modelling aspects of inelastic media* (pp. 217–226): Springer [Ed.].

Ostoja-Starzewski, M. (2006). Material spatial randomness: From statistical to representative volume element. *Probabilistic Engineering Mechanics, 21*, 112–132. https://doi.org/10.1016/j.probengmech.2005.07.007.

Otsuka, T., Brenner, R., & Bacroix, B. (2018). FFT-based modelling of transformation plasticity in polycrystalline materials during diffusive phase transformation. *International Journal of Engineering Science, 127*, 92–113.

Ozturk, D., Shahba, A., & Ghosh, S. (2016). Crystal plasticity FE study of the effect of thermo-mechanical loading on fatigue crack nucleation in titanium alloys. *Fatigue and Fracture of Engineering Materials and Structures, 39*, 752–769.

Panteghihi, A., & Bardella, L. (2016). On the finite element implementation of higher-order gradient plasticity, with focus on theories based on plastic distortion incompatibility. *Computer Methods in Applied Mechanics and Engineering*, *310*, 810–865.

Parton, V. Z., & Buryachenko, V. A. (1990). Stress fluctuations in elastic composites. *Soviet Physics Doklady*, *35*, 191–193.

Patel, D. K., & Kalidindi, S. R. (2017). Estimating the slip resistance from spherical nanoindentation and orientation measurements in polycrystalline samples of cubic metals. *International Journal of Plasticity*, *92*, 19–30.

Pei, L., Yun-Chang, X., & Qing, L. (2011). Plastic anisotropy and fracture behaviour of AZ31 magnesium alloy. *Transactions of Nonferrous Metals Society of China*, *21*, 880–884.

Peierls, R. (1940). The size of a dislocation. *Proceedings of the Physical Society*, *52*, 34.

Peirce, D. (1983). Shear band bifurcations in ductile single crystals. *Journal of the Mechanics and Physics of Solids*, *31*, 133–153.

Peirce, D., Asaro, R. J., & Needleman, A. (1982). An analysis of nonuniform and localized deformation in ductile single crystals. *Acta Metallurgica*, *30*, 1087–1119.

Peirce, D., Asaro, R. J., & Needleman, A. (1983). Material rate dependence and localized deformation in crystalline solids. *Acta Metallurgica*, *31*, 1951–1976.

Peric, D., de Souza Neto, E. A., Feijóo, R. A., Partovi, M., & Molina, A. J. C. (2011). On micro-to-macro transitions for multi-scale analysis of non-linear heterogeneous materials: Unified variational basis and finite element implementation. *International Journal for Numerical Methods in Engineering*, *87*, 149–170.

Petch, N. J. (1953). The cleavage strength of polycrystals. *Journal of the Iron and Steel Institute*, *174*, 25–28.

Pineau, A., McDowell, D. L., Busso, E. P., & Antolovich, S. D. (2016). Failure of metals II: Fatigue. *Acta Materialia*, *107*, 484–507.

Ponte Castañeda, P. (2002). Second-order homogenization estimates for nonlinear composites incorporating field fluctuations: I—Theory. *Journal of the Mechanics and Physics of Solids*, *50*, 737–757.

Ponte Castañeda, P. (1991). The effective mechanical properties of non-linear composites. *Journal of the Mechanics and Physics of Solids*, *37*, 45–71.

Ponte Castañeda, P. (1996). Exact second-order estimates for the effective mechanical properties of nonlinear composite materials. *Journal of the Mechanics and Physics of Solids*, *44*, 1757–1788.

Proudhon, H., Li, J., Reischig, P., Gueninchault, N., Forest, S., & Ludwig, W. (2015). Coupling diffraction contrast tomography with the finite element method. *Advanced Engineering Materials*, *18*, 903–912.

Proust, G., Tomé, C., & Kaschner, G. C. (2007). Modeling texture, twinning and hardening evolution during deformation of hexagonal materials. *Acta Materialia*, *55*, 2137–2148.

Przybyla, C. P., & McDowell, D. L. (2010). Microstructure-sensitive extreme value probabilities for high cycle fatigue of Ni-base superalloy IN100. *International Journal of Plasticity*, *26*, 372–394.

Przybyla, C. P., & McDowell, D. L. (2012). Microstructure-sensitive extreme-value probabilities of high-cycle fatigue for surface vs. subsurface crack formation in duplex Ti-6Al-4V. *Acta Materialia*, *60*, 293–305.

Przybyla, C. P., Musinski, W. D., Castelluccio, G. M., & McDowell, D. L. (2013). Microstructure-sensitive HCF and VHCF simulations. *International Journal of Fatigue*, *57*, 9–27.

Pushkareva, M., Adrien, J., Maire, E., Segurado, J., Llorca, J., & Weck, A. (2016). Three-dimensional investigation of grain orientation effects on void growth in commercially pure titanium. *Materials Science and Engineering: A*, *671*, 221–232.

Raabe, D., Sachtleber, M., Zhao, Z., Roters, F., & Zaefferer, S. (2001). Micromechanical and macromechanical effects in grain scale polycrystal plasticity experimentation and simulation. *Acta Materialia*, *49*, 3433–3441.

Radavich, J. F. (1989). The physical metallurgy of cast and wrought alloy 718. In E. A. Loria (Ed.), *Superalloy 718-metallurgy and applications* (pp. 229–240): TMS [Ed.].

Redenbach, C., Shklyar, I., & Andrä, H. (2012). Laguerre tessellations for elastic stiffness simulations of closed foams with strongly varying cell sizes. *International Journal of Engineering Science, 50*, 70–78.

Reina, C., & Conti, S. (2014). Kinematic description of crystal plasticity in the finite kinematic framework: A micromechanical understanding of F=FeFp. *Journal of the Mechanics and Physics of Solids, 67*, 40–61.

Reina, C., Schlmerkemper, A., & Conti, S. (2016). Derivation of F=FeFp as the continuum limit of crystalline slip. *Journal of the Mechanics and Physics of Solids, 89*, 231–254.

Rice, J. R. (1971). Inelastic constitutive relations for solids: An internal-variable theory and its application to metal plasticity. *Journal of the Mechanics and Physics of Solids, 19*, 433–455.

Richards, A. W., Lebensohn, R. A., & Bhattacharya, K. (2013). Interplay of martensitic phase transformation and plastic slip in polycrystals. *Acta Materialia, 61*, 4384–4397.

Rodríguez-Galán, D., Sabirov, I., & Segurado, J. (2015). Temperature and stain rate effect on the deformation of nanostructured pure titanium. *International Journal of Plasticity, 70*, 191–205.

Rodríguez-Galán, D., Segurado, J., & Romero, I. (2017). Working document.

Roters, F., Eisenlohr, P., Bieler, T. R., & Raabe, D. (2010). *Crystal plasticity finite element methods: In materials science and engineering*. Wiley-VCH.

Roters, F., Eisenlohr, P., Hantcherli, L., Tjahjanto, D. D., Bieler, T. R., & Raabe, D. (2010). Overview of constitutive laws, kinematics, homogenization and multiscale methods in crystal plasticity finite-element modeling: Theory, experiments, applications. *Acta Materialia, 58*, 1152–1211.

Sachs, G. (1928). On the derivation of a condition of flowing. *Verein Deutscher Ingenieure, 72*, 734–736.

Sánchez-Martín, R., Pérez-Prado, M. T., Segurado, J., Bohlen, J., Gutiérrez-Urrutia, I., LLorca, J., & Molina-Aldareguia, J. M. (2014a). Measuring the critical resolved shear stresses in Mg alloys by instrumented nanoindentation. *Acta Materialia, 71*, 283–292.

Sánchez-Martín, R., Pérez-Prado, M. T., Segurado, J., Bohlen, J., Gutiérrez-Urrutia, I., LLorca, J., & Molina-Aldareguia, J. M. (2014b). Measuring the critical resolved shear stresses in Mg alloys by instrumented nanoindentation. *Acta Materialia, 71*, 283–292.

Sangid, M. D. (2013). The physics of fatigue crack initiation. *International Journal of Fatigue, 57*, 58–72.

Sangid, M. D., Maier, H. J., & Sehitoglu, H. (2011). An energy-based microstructure model to account for fatigue scatter in polycrystals. *Journal of the Mechanics and Physics of Solids, 59*, 595–609.

Schmidt, W. L. S., Lauridsen, E. M., & Poulsen, H. F. (2008). X-ray diffraction contrast tomography: A novel technique for three-dimensional grain mapping of polycrystals. 1. Direct beam case. *Journal of Applied Crystallography, 41*, 302–309.

Segurado, J., González, C., & LLorca, J. (2003). A numerical investigation of the effect of particle clustering on the mechanical properties of composites. *Acta Materialia, 51*, 2355–2369.

Segurado, J., Lebensohn, R. A., Llorca, J., & Tomé, C. (2012). Multiscale modeling of plasticity based on embedding the viscoplastic self-consistent formulation in implicit finite elements. *International Journal of Plasticity, 28*, 124–140.

Segurado, J., & LLorca, J. (2002). A numerical approximation to the elastic properties of sphere-reinforced composites. *Journal of the Mechanics and Physics of Solids, 50*, 2107–2121.

Segurado, J., & LLorca, J. (2005). Computational micromechanics of composites: The effect of particle spatial distribution. *Mechanics of Materials, 38*, 873–883.

Segurado, J., & LLorca, J. (2013). Simulation of the deformation of polycrystalline nanostructured Ti by computational homogenization. *Computational Materials Science, 76*, 3–11.

Sencer, B. H., Maloy, S. A., & Gray, G. T. (2005). The influence of shock-pulse shape on the structure/property behavior of copper and 316L austenitic stainless steel. *Acta Materialia*, *53*, 3293–3303.

Shahba, A., & Ghosh, S. (2016). Crystal plasticity FE modeling of Ti alloys for a range of strain-rates. Part I: A unified constitutive model and flow rule. *International Journal of Plasticity*, *87*, 48–68.

Shenoy, M. (2006). *Constitutive modeling and life prediction in ni-base superalloys*. (Ph.D. thesis). Georgia Institute of Technology.

Shenoy, M., Zhang, J., & McDowell, D. L. (2007). Estimating fatigue sensitivity to polycrystalline Ni-base superalloy microstructures using a computational approach. *Fatigue and Fracture of Engineering Materials and Structures*, *30*, 889–904.

Shu, J. Y., & Fleck, N. A. (1999). Strain gradient crystal plasticity: Size-dependent deformation of bicrystals. *Journal of the Mechanics and Physics of Solids*, *47*, 297–324.

Song, Y. S., Lee, M. R., & Kim, J. T. (2005). Effect of grain size for the tensile strength and the low cycle fatigue at elevated temperature of alloy 718 cogged by open die forging press. In E. A. Loria (Ed.), *Superalloys 718, 625, 706 and derivatives* (pp. 539–549): TMS [Ed.].

Spowart, J. E. (2006). Automated serial sectioning for 3-D analysis of microstructures. *Scripta Materialia*, *55*, 5–10.

Spowart, J. E., Mullens, H. M., & Puchala, B. T. (2003). Collecting and analyzing microstructures in three dimensions: A fully automated approach. *The Journal of The Minerals, Metals & Materials Society*, *55*, 35–37.

Staroselsky, A., & Anand, L. (2003). A constitutive model for hcp materials deforming by slip and twinning: Application to magnesium alloy AZ31B. *International Journal of Plasticity*, *19*, 1843–1864.

Stelmashenko, N. A., Walls, M. G., Brown, L. M., & Milman, Y. (1993). Microindentations on W and Mo oriented single crystals: An STM study. *Acta Metallurgica et Materialia*, *41*, 2855–2865.

Stinville, J. C., Lenthe, W. C., Miao, J., & Pollock, T. M. (2016). A combined grain scale elasticplastic criterion for identification of fatigue crack initiation sites in a twin containing polycrystalline nickel-base superalloy. *Acta Materialia*, *103*, 461–473.

Stinville, J. C., Vanderesse, N., Bridier, F., Bocher, P., & Pollock, T. M. (2015). High resolution mapping of strain localization near twin boundaries in a nickel-based superalloy. *Acta Materialia*, *98*, 29–42.

Stukowski, A., Cereceda, D., Swinburne, T. D., & Marian, J. (2015). Thermally-activated non-Schmid glide of screw dislocations in W using atomistically-informed kinetic Monte Carlo simulations. *International Journal of Plasticity*, *65*, 108–130.

Su, Y., Zambaldi, C., Mercier, D., Eisenlohr, P., Bieler, T. R., & Crimp, M. A. (2016). Quantifying deformation processes near grain boundaries in α titanium using nanoindentation and crystal plasticity modeling. *International Journal of Plasticity*, *86*, 170–186.

Sulsky, D., Zhou, S. J., & Schreyer, H. L. (1995). Application of a particle-in-cell method to solid mechanics. *Computational Physics Communications*, *87*, 236–252.

Suresh, S. (2012). *Fatigue of materials*. Cambridge University Press.

Sweeney, C. A., McHugh, P. E., McGarry, J. P., & Leen, S. B. (2012). Micromechanical methodology for fatigue in cardiovascular stents. *International Journal of Fatigue*, *44*, 202–216.

Sweeney, C. A., O'Brien, B., Dunne, F. P. E., McHugh, P. E., & Leen, S. B. (2014). Strain-gradient modelling of grain size effects on fatigue of CoCr alloy. *Acta Materialia*, *78*, 341–353.

Sweeney, C. A., O'Brien, B., Dunne, F. P. E., McHugh, P. E., & Leen, S. B. (2015). Micro-scale testing and micromechanical modelling for high cycle fatigue of CoCr stent material. *Journal of the Mechanical Behavior of Biomedical Materials*, *46*, 244–260.

Sweeney, C. A., O'Brien, B., McHugh, P. E., & Leen, S. B. (2014). Experimental characterisation for micromechanical modelling of CoCr stent fatigue. *Biomaterials*, *35*, 36–48.
Sweeney, C. A., Vorster, W., Leen, S. B., Sakurada, E., McHugh, P. E., & Dunne, F. P. E. (2013). The role of elastic anisotropy, length scale and crystallographic slip in fatigue crack nucleation. *Journal of the Mechanics and Physics of Solids*, *61*, 1224–1240.
Tanaka, K., & Mura, T. (1981). A dislocation model for fatigue crack initiation. *ASME Journal of Applied Mechanics*, *48*, 97–103.
Tandon, G., & Weng, G. (1988). A theory of particle-reinforced plasticity. *Journal of Applied Mechanics*, *55*, 126–135.
Taylor, G. I. (1938). Plastic strain in metals. *Journal of the Institute of Metals*, *62*, 307–324.
Taylor, G. I., & Elam, C. F. (1923). The distortion of an aluminium crystal during a tensile test. *Proceedings of the Royal Society of London A: Mathematical, Physical and Engineering Sciences*, *102*, 643–667.
Taylor, G. I., & Elam, C. F. (1925). The plastic extension and fracture of aluminium crystals. *Proceedings of the Royal Society of London A: Mathematical, Physical and Engineering Sciences*, *108*, 28–51.
Thomason, P. F. (1990). *Ductile fracture of metals*. Pergamon Press.
Tomé, C. (1999). Self-consistent polycrystal models: A directional compliance criterion to describe grain interactions. *Modelling and Simulation in Material Science and Engineering*, 7, 723–728.
Tomé, C., Canova, G. R., Kocks, U. F., Christodoulou, N., & Jonas, J. J. (1984). The relation between macroscopic and microscopic strain hardening in F.C.C. polycrystals. *Acta Metallurgica*, *32*(10), 1637–1653.
Tomé, C. N., Agnew, S. R., & Yoo, M. H. (2001). Application of texture simulation to understanding mechanical behavior of Mg and solid solution alloy containing Li or Y. *Acta Materialia*, *49*, 4277–4289.
Tomé, C. N., Lebensohn, R. A., & Necker, C. T. (2002). Orientation correlations and anisotropy of recrystallized aluminum. *Metallurgical and Materials Transactions A*, *33*, 2635–2648.
Tomé, C. N., Lebensohn, R. A., & Kocks, U. F. (1991). A model for texture development dominated by deformation twinning: Application to zirconium alloys. *Acta Metallurgica et Materialia*, *39*(11), 2667–2680.
Tomé, C. N., Maudlin, P. J., Lebensohn, R. A., & Kaschner, G. C. (2001). Mechanical response of zirconium: I. Derivation of a polycrystal constitutive law and finite element analysis. *Acta Materialia*, *49*, 3085–3096.
Torquato, S. (2001). *Random heterogeneous materials*. New York: Springer.
Totry, E., González, C., & LLorca, J. (2008). Failure locus of fiber-reinforced composites under transverse compression and out-of-plane shear. *Composites Science and Technology*, *68*, 829–839.
Tucker, J. C., Cerrone, A. R., III, Ingraffea, A. R., & Rollett, A. D. (2015). Crystal plasticity finite element analysis for Ren88DT statistical volume element generation. *Modelling and Simulation in Materials Science and Engineering*, *23*, 035003.
Uchic, M. D., Groeber, M., Wheeler, R., Scheltens, F., & Dimiduk, D. M. (2004). Augmenting the 3D characterization capability of the dual beam FIB-SEM. *Microscopy and Microanalysis*, *10*, 1136–1137.
Uchic, M. D., Groeber, M. A., Dimiduk, D. M., & Simmons, J. P. (2006). 3D microstructural characterization on nickel superalloys via serial-sectioning using a dual-beam FIB-SEM. *Scripta Materialia*, *55*, 23–28.
Upadhyay, M. V., Capolungo, L., Taupin, V., Fressengeas, C., & Lebensohn, R. A. (2016). A higher order elasto-viscoplastic model using fast Fourier transforms: Effects of lattice curvatures on mechanical response of nanocrystalline metals. *International Journal of Plasticity*, *83*, 126–152.

Vachhani, S. J., & Kalidindi, S. R. (2015). Grain-scale measurement of slip resistances in aluminum polycrystals using spherical nanoindentation. *Acta Materialia, 60*, 27–36.
Van Houtte, P., Kanjarla, A. K., Van Bael, A., Seefeldt, M., & Delannay, L. (2006). Multiscale modelling of the plastic anisotropy and deformation texture of polycrystalline materials. *European Journal of Mechanics A/Solids, 25*, 634–648.
Van Houtte, P., Yerra, S. K., & Van Bael, A. (2009). The Facet method: A hierarchical multilevel modelling scheme for anisotropic convex plastic potentials. *International Journal of Plasticity, 25*, 332–360.
Venkatramani, G., Ghosh, S., & Mills, M. (2007). A size-dependent crystal plasticity finite-element model for creep and load shedding in polycrystalline titanium alloys. *Acta Materialia, 55*, 3971–3986.
Vondrejc, J., Zeman, J., & Marek, I. (2014). An FFT-based Galerkin method for homogenization of periodic media. *Computers and Mathematics with Applications, 68*, 156–173.
Vondrous, A., Bienger, P., Schreijäg, S., Selzer, M., Schneider, D., Nestler, B., … Mönig, R. (2015). Combined crystal plasticity and phase-field method for recrystallization in a process chain of sheet metal production. *Computational Mechanics, 55*, 439–452.
Wan, V. V. C., Jiang, J., MacLachlan, D. W., & Dunne, F. P. E. (2016). Microstructure-sensitive fatigue crack nucleation in a polycrystalline Ni superalloy. *International Journal of Fatigue, 90*, 181–190.
Wan, V. V. C., MacLachlan, D. W., & Dunne, F. P. E. (2016). Integrated experiment and modelling of microstructurally-sensitive crack growth. *International Journal of Fatigue, 91*, 110–123.
Wang, F., Sandlobes, S., Diehl, M., Sharma, L., Roters, F., & Raabe, D. (2014). In situ observation of collective grain-scale mechanics in Mg and Mg-rare earth alloys. *Acta Materialia, 80*, 77–93.
Wen, W., Borodachenkova, M., Tomé, C. N., Vincze, G., Rauch, E. F., Barlat, F., & Gracio, J. J. (2016). Mechanical behavior of Mg subjected to strain path changes: Experiments and modeling. *International Journal of Plasticity, 73*, 171–183.
Wenk, H. R., Canova, G. R., Brechet, Y., & Flandin, L. (1997). A deformation-based model for recrystallization of anisotropic materials. *Acta Materialia, 45*, 3283–3296.
Werwer, M., & Cornec, A. (2000). Numerical simulation of plastic deformation and fracture in polysynthetically twinned (PST) crystals of TiAl. *Computational Materials Science, 19*, 97–107.
Willis, J. R. (1981). Variational and related methods for the overall properties of composites. *Advanced Applied Mechanics, 21*, 1–78.
Willot, F. (2015). Fourier-based schemes for computing the mechanical response of composites with accurate local fields. *Comptes Rendus Mécanique, 34*, 232–245.
Wu, P. D., Tomé, C. N., Agnew, S. R., Raeisinia, B., & Wang, H. (2010). Evaluation of self consistent polycrystal plasticity models for magnesium alloy AZ31B sheet. *International Journal of Solids and Structures, 47*, 2905–2917.
Xue, X., Righetti, F., Telley, H., Liebling, T. M., & Mocellin, A. (1997). The Laguerre model for grain growth in three dimensions. *Philosophical Magazine Part B, 75*, 567–585.
Yasi, J. A., Nogaret, T., Trinkle, D. R., Qi, Y., Hector, L. G., Jr, & Curtin, W. A. (2009). Basal and prism dislocation cores in magnesium: Comparison of first-principles and embedded-atom-potential methods predictions. *Modelling and Simulation in Materials Science and Engineering, 17*, 055012.
Yefimov, S., van der Giessen, E., & Groma, I. (2004). Bending of a single crystal: Discrete dislocation and nonlocal crystal plasticity simulations. *Modelling and Simulation in Materials Science and Engineering, 12*, 1069–1086.
Yeratapally, S. R., Glavicic, M. G., Hardy, M., & Sangid, M. D. (2016). Microstructure based fatigue life prediction framework for polycrystalline nickel-base superalloys with

emphasis on the role played by twin boundaries in crack initiation. *Acta Materialia, 107*, 152–167.
Zambaldi, C., Yang, Y., Bieler, T. R., & Raabe, D. (2012). Orientation informed nanoindentation of α-titanium: Indentation pileup in hexagonal metals deforming by prismatic slip. *Journal of Materials Research, 27*, 356–367.
Zambaldi, C., Zehnder, C., & Raabe, D. (2015). Orientation dependent deformation by slip and twinning in magnesium during single crystal indentation. *Acta Materialia, 91*, 267–288.
Zecevic, M., Beyerlein, I. J., & Knezevic, M. (2017). Coupling elasto-plastic self-consistent crystal plasticity and implicit finite elements: Applications to compression, cyclic tension-compression, and bending to large strains. *International Journal of Plasticity, 93*, 187–211.
Zecevic, M., Pantleon, W., Lebensohn, R. A., McCabe, R. J., & Knezevic, M. (2017). Predicting intragranular misorientation distributions in polycrystalline metals using the viscoplastic self-consistent formulation. *Acta Materialia, 140*, 398–410.
Zeman, J., de Geus, T. W. J., Vondrejc, J., Peerlings, R. H. J., & Geers, M. G. D. (2017). A finite element perspective on nonlinear FFT-based micromechanical simulations. *International Journal for Numerical Methods in Engineering, 110*, 903–926.
Zeman, J., Vondřejc, J., Novák, J., & Marek, I. (2010). Accelerating a FFT-based solver for numerical homogenization of periodic media by conjugate gradients. *Journal of Computational Physics, 229*(21), 8065–8071.
Zhang, C., Li, H., Eisenlohr, P., Liu, W., Boehlert, C. J., Crimp, M. A., & Bieler, T. R. (2015). Effect of realistic 3D microstructure in crystal plasticity finite element analysis of polycrystalline Ti-5Al-2.5Sn. *International Journal of Plasticity, 69*, 21–35.
Zhang, J., & Joshi, S. P. (2012). Phenomenological crystal plasticity modeling an detailed micromechanical investigations of pure magnesium. *Journal of the Mechanics and Physics of Solids, 60*, 945–972.

CHAPTER TWO

Metadamping: Dissipation Emergence in Elastic Metamaterials

Clémence L. Bacquet*, Hasan Al Ba'ba'a†, Michael J. Frazier‡, Mostafa Nouh†, Mahmoud I. Hussein*[,1]

*Ann and H.J. Smead Department of Aerospace Engineering Sciences, University of Colorado Boulder, Boulder, CO, United States

†Department of Mechanical and Aerospace Engineering, State University of New York at Buffalo, Buffalo, NY, United States

‡Department of Mechanical and Aerospace Engineering, University of California San Diego, San Diego, CA, United States

[1]Corresponding author: e-mail address: mih@colorado.edu

Contents

Advances in Applied Mechanics, Volume 51
ISSN 0065-2156
https://doi.org/10.1016/bs.aams.2018.09.001

Abstract

Resonant elastic metamaterials are artificial material systems that exhibit unique dynamical properties shaped by the intrinsic interaction between resonances and traveling dispersive waves. In this chapter, we provide a technical review of the recently proposed concept of dissipation emergence in elastic metamaterials. This concept is termed "metadamping." Unlike conventional materials used to dampen vibrations where the damping capacity is affected by the atomic configuration, defects, and/or rheological properties, here the level of dissipation is controlled via the dynamics of the metamaterial's resonant substructures. In this manner, it is possible to create a net material system that is both stiff and highly damped, to absorb vehicle vibrations for example. The chapter starts with a motivation and introduction of metadamping, and then presents an in-depth analysis and parametric study of metadamping in the context of both locally and nonlocally resonant elastic metamaterials modeled as mass-spring-dashpot systems. The effect of the core damping model (e.g., viscous vs nonviscous) is also examined. Finally, a review is given of metadamping in a pillared beam that has recently been investigated by experiments, simulations, and theory.

1. PHONONIC MATERIALS: DYNAMICAL DESIGN AT THE MATERIAL LEVEL

1.1 Material Dynamics

The observed properties of materials—acoustic, mechanical, etc.—are the cumulative result of many small-scale processes and interactions occurring within the material microstructure. Many technological applications require extraordinary materials that satisfy exceptional demands. Thus, the ability to tailor the microstructure and, consequently, the material behavior for applications is desirable. Chemistry represents a conventional avenue for manipulating the microstructure at the atomic/molecular/grain-boundary scale to achieve the desired material performance. Alternatively, especially with the advent of additive manufacturing technology, the internal structure may be designed and implemented at the *macroscopic* level to elicit unique, even counterintuitive, performance from the resulting architected material—with the possibility of extending the frontier of material behavior. Here, we focus on the dynamical aspects of such materials, termed *phononic materials*, which include phononic crystals (PC) and acoustic/elastic metamaterials (Kushwaha, Halevi, Dobrzynski, & Djafari-Rouhani, 1993; Liu et al., 2000; Sigalas & Economou, 1992). And in particular, we focus on the recently proposed concept of *metadamping* which allows us to design the material architecture in a way that renders it more dissipative than the nominal case without such architecture (Hussein & Frazier, 2013b).

To confine the discussion to within a manageable scope, we examine here only a sample of their physics and potential utility.

Phononic materials, generally, emerge from the periodic arrangement of small-scale building blocks—reminiscent of molecules in natural crystals—which, through scattering/interference and/or resonance phenomena, act to control the propagation of acoustic/elastic plane waves. Fundamentally, over specific frequency ranges, i.e., the *pass bands*, a phononic material is transparent to vibrational plane waves, which propagate at different speeds in contrast to the constant sonic speeds of many nondispersive conventional materials. Outside these frequency bands, i.e., within the *stop bands* or *band gaps*, the material internal structure scatters and/or localizes the wave energy, prohibiting transmission in all or specific directions. Waveguides and acoustic filters are two basic applications of this behavior. However, through careful design of building blocks forming the material's internal structure, the unique dynamics of phononic materials may be exploited in a myriad of original applications [e.g., flat acoustic lenses (Yang et al., 2004; Zhang & Liu, 2004), acoustic/elastic cloaks (Brun, Guenneau, & Movchan, 2009; Cummer & Schurig, 2007), phononic subsurfaces for flow control (Hussein, Biringen, Bilal, & Kucala, 2015), to name a few] (Deymier, 2013; Hussein, Leamy, & Ruzzene, 2014; Khelif & Adibi, 2015; Laude, 2015; Phani & Hussein, 2017).

One of the earliest and simplest examples of a phononic material is the one-dimensional medium composed of periodically alternating layers of differing composition—mass-density and elasticity—such as shown in Fig. 1A (Kohn, Krumhansl, & Lee, 1972; Lee & Yang, 1973; Sun, Achenbach, & Herrmann, 1968). Within a homogeneous medium, a vibrational wave of arbitrary frequency ω, and wavenumber (i.e., spatial frequency) κ, propagates at a constant speed, $c_0 = \omega/\kappa$. As can be seen in Fig. 1A, within a phononic material, waves of different frequencies travel at different speeds—a phenomenon referred to as *dispersion*—indicative of a nonlinear relationship between ω and κ. For the layered medium, this is accommodated by the superposition of waves transmitted and reflected at the layer interfaces. Moreover, over specific frequency ranges, the scattering/interference is sufficient to prevent the associated waves from propagating within the material. Fig. 1C illustrates each of these scenarios where the displacement profiles are plotted over several unit cells, the fundamental repeating structure. Within the pass band, the wave propagates through the phononic material such that the displacement profile encompasses the entire–theoretically infinite—domain. In contrast, within the band

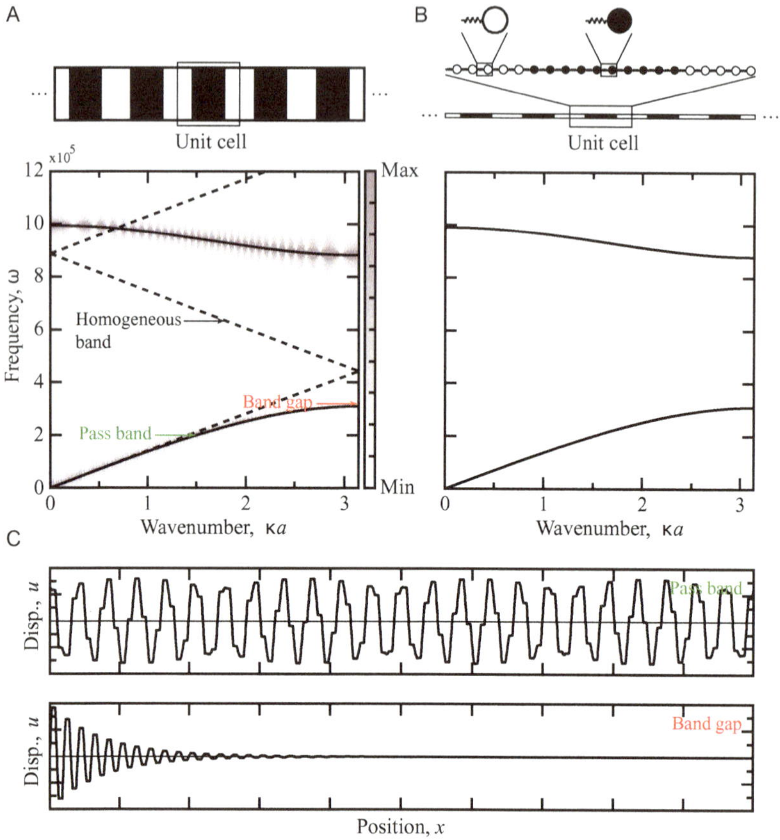

Fig. 1 (A) Dispersion diagram of a layered medium (continuous model) as determined from numerical and analytical treatments. (B) Dispersion diagram of a mass-spring system (discrete model) determined analytically. (C) Displacement profile in corresponding finite layered medium at different frequencies illustrating the contrast in response between the pass band (*top*) and band gap (*bottom*) cases.

gap, the wave amplitude decays with distance and the displacement nodes remain fixed in space for all time.

In Fig. 1A, we present the results of simulation and analytics. For the simulation, a layered medium composed of 100 cells is disturbed by an impulse at the mid-span, producing a wide spectrum of waves that propagate away from the site of initial disturbance. Prior to reaching either end of the medium, the waves interact only with each other and the underlying material periodicity of the system, mimicking an infinite medium. Under these conditions, a Fourier transform of the displacement in space (with $e^{i\kappa x}$) and time (with $e^{-i\omega t}$) produces the density plot shown, where the narrow, darker regions identify waves with spectral characteristics (i.e., ω and κ) suitable for transmission through the layered medium and the brighter regions represent incompatible characteristics. The color map in Fig. 1A and similar diagrams produced via simulation are instructive in revealing the dispersion properties of the relevant phononic material design but are computationally expensive, especially for two- and three-dimensional systems with complex unit-cell designs. Alternatively, the periodic arrangement of the material constituents, reminiscent of molecules in natural crystals, opens phononic materials to theoretical analysis via Floquet–Bloch theorem (henceforth simplified to Bloch) from solid-state physics. For one-dimensional elastodynamic problem, this is expressed as (Bloch, 1929; Floquet, 1883):

$$u(x,t) = \tilde{u}(x)\mathrm{e}^{\mathrm{i}(\kappa x - \omega t)} \tag{1}$$

where $u(x, t)$ is the displacement. The amplitude function $\tilde{u}(x)$ has the same periodicity at the underlying medium, $x \in [0, \pm a]$ where a is the lattice spacing or unit-cell length. As a consequence, $u(x \pm na, t) = u(x, t)\mathrm{e}^{\pm \mathrm{i}\kappa na}$ for any arbitrary integer n, which permits the solution for $x \in (-\infty, \infty)$ to be determined from the analysis of a finite segment of material of dimension a—the unit cell. This modeling concept applies to two- and three-dimensional systems as well. By considering only a single unit cell rather than the potential myriad of cells, Bloch's theorem greatly reduces computation demands and yields a compact, theoretical representation of the wave propagation characteristics. Fig. 1A illustrates excellent agreement between numerical (density plot) and analytical (smooth curves) results for the layered medium.

As mentioned previously, the layered medium is one of the most elementary practical representations of a phononic material which could serve as a simple demonstration. However, in what follows, we utilize simple lumped-parameter models that capture all the essential physics and make the connection between unit-cell configuration and performance more accessible. Indeed, in regard to the ability to capture the essential physics of a problem, as shown in Fig. 1B, the dispersion diagram of the corresponding lumped-parameter model closely resembles that of its continuum counterpart shown in Fig. 1A. This model uses masses and springs to represent each layer of the unit cell, with each mass and spring defined as $m = \rho AL$ and $k = EA/L$, respectively, where ρ is the density, E is the Young's modulus, and A and L denote the rod cross-sectional area and length, respectively.

1.2 Consideration of Material Damping

Inherent to every material, damping mechanisms (e.g., friction) dissipate the wave energy, affecting the propagation and spatial attenuation characteristics

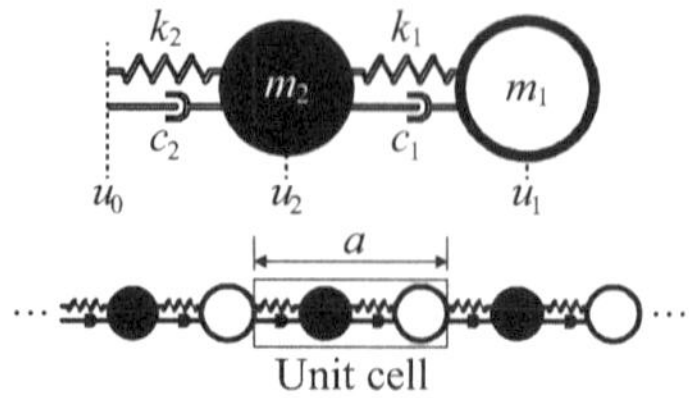

Fig. 2 Discrete model of a phononic material constructed from a unit cell of two interacting masses which repeats infinitely along one dimension.

(Crandall, 1970). However, damping is often unaccounted for in theoretical analysis, removing a practical aspect of material performance and a mean to control that performance. In the following, we discuss the physics and utility of damping in phononic materials [for further reading, see, e.g., Hussein (2009b), Hussein and Frazier (2010), Frazier and Hussein (2016), Al Ba'ba'a and Nouh (2017)]. To this end, we primarily use the mass-spring-damper model shown in Fig. 2, which forms the basis of one-dimensional phononic material similar to the discrete example discussed previously.

The set of equations that describe the motion of each degree of freedom in the unit cell are collected in matrix form as [see Hussein and Frazier (2013a) and Frazier and Hussein (2016) for the derivations]

$$\mathbf{M}\ddot{\mathbf{u}} + \mathbf{C}\dot{\mathbf{u}} + \mathbf{K}\mathbf{u} = \mathbf{f} \tag{2}$$

where the mass $\mathbf{M}$, viscous damping $\mathbf{C}$, and stiffness $\mathbf{K}$ matrices are, respectively, given by

$$\mathbf{M} = m_2 \begin{bmatrix} 1/r_m & 0 & 0 \\ 0 & 1 & 0 \\ 0 & 0 & 0 \end{bmatrix} \tag{3a}$$

$$\mathbf{C} = c_2 \begin{bmatrix} 1/r_c & -1/r_c & 0 \\ -1/r_c & 1/r_c + 1 & -1 \\ 0 & -1 & 1 \end{bmatrix} \tag{3b}$$

$$\mathbf{K} = k_2 \begin{bmatrix} 1/r_k & -1/r_k & 0 \\ -1/r_k & 1/r_k + 1 & -1 \\ 0 & -1 & 1 \end{bmatrix} \tag{3c}$$

consistent with the displacement degrees of freedom $\mathbf{u}^{\mathrm{T}} = [u_1\ u_2\ u_0]$. The superposed dot represents differentiation with respect to time. The forces applied by neighboring cells are organized in $\mathbf{f}^{\mathrm{T}} = [f_1\, f_2\, f_0]$. Here, we introduce the material property relations $r_m = m_2/m_1$, $r_c = c_2/c_1$, $r_k = k_2/k_1$. Note, Eq. (2) represents the equation of motion of a single unit cell viewed as a finite entity. To model the dynamics of the corresponding infinite, periodic medium, we apply Bloch boundary conditions which establish a relationship (perpetuating in space ad infinitum) between the unit-cell boundary states and those of its neighbors. We define a condensed displacement vector containing only the essential degrees of freedom, that is, the internal and essential boundary displacement, $\hat{\mathbf{u}}^{\mathrm{T}} = [u_1 \;\; u_2]$. From Bloch's theorem, $u_0 = u_1 \mathrm{e}^{-\mathrm{i}\kappa a}$. Thus, the condensed and full set of displacement degrees of freedom are related via the transformation, $\mathbf{T}$, as follows

$$\mathbf{u} = \mathbf{T}\hat{\mathbf{u}} = \begin{bmatrix} 1 & 0 \\ 0 & 1 \\ \mathrm{e}^{-\mathrm{i}\kappa a} & 0 \end{bmatrix} \begin{bmatrix} u_1 \\ u_2 \end{bmatrix}. \tag{4}$$

When applied to the equation of motion, the transformation yields

$$\hat{\mathbf{M}}\ddot{\hat{\mathbf{u}}} + \hat{\mathbf{C}}\dot{\hat{\mathbf{u}}} + \hat{\mathbf{K}}\hat{\mathbf{u}} = \mathbf{0} \tag{5}$$

with $\hat{\mathbf{M}} = \mathbf{T}^{\mathrm{H}}\mathbf{M}\mathbf{T}$, $\hat{\mathbf{C}} = \mathbf{T}^{\mathrm{H}}\mathbf{C}\mathbf{T}$, and $\hat{\mathbf{K}} = \mathbf{T}^{\mathrm{H}}\mathbf{K}\mathbf{T}$, where the superscript "H" represents the complex transpose operation. Given that Bloch analysis is concerned with free wave motion, we set $\mathbf{T}^{\mathrm{H}}\mathbf{f} = \mathbf{0}$.

From a theoretical perspective, the wave frequency is often assumed to be strictly real-valued regardless of the damping condition (Castanier & Pierre, 1993; Collet, Ouisse, Ruzzene, & Ichchou, 2011; Farzbod & Leamy, 2011; Langley, 1994a; Mead, 1973; Merheb et al., 2008; Moiseyenko & Laude, 2011). Consequently, according to Eq. (1), wave attenuation (due to interference and/or damping) may only occur over the propagation distance as the wavenumber is free to be complex-valued, $\kappa = \kappa_{\mathrm{R}} + \mathrm{i}\kappa_{\mathrm{I}}$. This may be considered the controlled laboratory perspective, that is, the material response is what may be expected when exciting an extended, but finite, structure at the boundary at a prescribed frequency; hence supplying continuous injection of wave energy. However, if we are interested in the inherent material response following an initial input of energy, our analysis of wave propagation in damped phononic materials allows the temporal frequency to be complex. In this context, waves attenuate in space (due to scattering/interference and resonance) and time (due to material damping). This may be considered the free-motion perspective as the dynamics stem purely from the response of the material to an initial disturbance, as opposed to a sustained external stimulation. In the absence of damping, the two approaches are equivalent. Damping reveals the unique dynamics of each operation condition, and in principle it is possible to solve for an all-complex band structure where both the frequencies and wavenumbers are complex (Frazier & Hussein, 2016; Hussein, Frazier, & Abedinnassab, 2013). For real-valued frequencies, the band-gap effect is unadulterated only in the complete absence of damping (not seen in practice), otherwise, the band gap closes with the slightest level of damping. Conversely, for complex-valued frequencies, the impact of wave interferences is more intact and the band gap remains complete except under extreme conditions (Frazier & Hussein, 2016; Phani & Hussein, 2013).

For general free wave motion, the displacement solution varies according to $e^{\lambda t}$ where λ is the complex frequency whose real and imaginary components are to be determined. Modifying Bloch's theorem accordingly and subsequently applying the derivatives in Eq. (2) yields the following quadratic eigenvalue problem in λ

$$(\lambda^2\hat{\mathbf{M}} + \lambda\hat{\mathbf{C}} + \hat{\mathbf{K}})\hat{\mathbf{u}} = \mathbf{0} \tag{6}$$

Here we can utilize the state-space transformation from structural dynamics (Wagner & Adhikari, 2003) to formulate an eigenvalue problem with twice the degrees of freedom but linear in construction (Frazier & Hussein, 2015; Hussein & Frazier, 2010, 2013a). We define the state vector, $\hat{\mathbf{y}}^{\mathrm{T}} = [\dot{\hat{\mathbf{u}}} \;\; \hat{\mathbf{u}}]$, enabling the state-space transformation

$$\begin{bmatrix} \mathbf{0} & \hat{\mathbf{M}} \\ \hat{\mathbf{M}} & \hat{\mathbf{C}} \end{bmatrix} \dot{\hat{\mathbf{y}}} + \begin{bmatrix} -\hat{\mathbf{M}} & \mathbf{0} \\ \mathbf{0} & \hat{\mathbf{K}} \end{bmatrix} \hat{\mathbf{y}} = \mathbf{0} \tag{7}$$

Assuming the time-dependence of the state vector is also of the form $e^{\lambda t}$, we formulate the following generalized linear eigenvalue problem in λ

$$\left(\lambda \begin{bmatrix} \mathbf{0} & \hat{\mathbf{M}} \\ \hat{\mathbf{M}} & \hat{\mathbf{C}} \end{bmatrix} + \begin{bmatrix} -\hat{\mathbf{M}} & \mathbf{0} \\ \mathbf{0} & \hat{\mathbf{K}} \end{bmatrix}\right) \hat{\mathbf{y}} = \mathbf{0} \tag{8}$$

In general, the complex eigenvalues take the form

$$\lambda(\kappa) = -\xi(\kappa)\omega_{\mathrm{r}}(\kappa) \pm \mathrm{i}\omega_{\mathrm{d}}(\kappa) \tag{9}$$

Inserting Eq. (9) into $e^{\lambda t}$, it is apparent that $\omega_{\mathrm{d}}(\kappa)$, the damped wave frequency, leads to temporal oscillations and $\xi(\kappa)\omega_{\mathrm{r}}(\kappa)$ is responsible for the decay of the wave amplitude over time. Specifically, the quantity $\xi(\kappa)$ is the dimensionless damping ratio (loss factor) and $\omega_{\mathrm{r}}(\kappa)$ is termed the "resonant frequency." The wavenumber-dependent damping ratio relation is extracted from the complex eigenvalues as follows

$$\xi(\kappa) = -\frac{\mathrm{Re}[\lambda(\kappa)]}{|\lambda(\kappa)|} \tag{10}$$

Naturally, in the absence of damping, $\xi(\kappa) = 0$ and $\omega_d(\kappa)$ is identical to the undamped solution $\omega(\kappa)$. In addition, due to the method of extraction, the maximum value that $\xi(\kappa)$ may attain over a range of wavenumbers is unity, which requires that $\omega_d(\kappa) = 0$ over the corresponding κ range.

Fig. 3A and B shows the damped frequency and damping ratio dispersion diagrams for specific normalized damping intensities β/ω_0, including the undamped scenario $\beta/\omega_0 = 0$ for reference. The ratio $\beta = c_2/m_2$ is a measure of the damping intensity and $\omega_0^2 = k_2/m_2$ is the natural frequency of the oscillator m_2. The specific material property ratios selected for this example are $r_m = 3$, $r_c = 1/2$, $r_k = 1$, and $\omega_0 = 65.32$ rad/s. Notice that viscous damping compacts the frequency band structure, affecting higher frequencies more acutely than lower ones. This is corroborated by the corresponding damping ratios in Fig. 3B which expand as the damping intensity increases, most rapidly for wavenumbers corresponding to higher frequencies. In the extreme, we can observe two intriguing phenomena not encountered in real-frequency analysis: band overtaking and band cut-off (Hussein, 2009b; Hussein & Frazier, 2010; Phani & Hussein, 2013). Band overtaking occurs when frequency bands cross; band cut-off describes the scenario in which frequency bands do not span the full range of wavenumbers. In Fig. 3A, we observe a case of band overtaking at $\beta/\omega_0 = 0.45$. In particular, we observe that the upper band (dashed) drops at a much faster rate than the lower band (solid) as the corresponding damping ratio is greater, enabling the overtake. This is quantitatively illustrated by tracking the band-gap width as a function of damping intensity, as illustrated in Fig. 3C. In some cases, if the overtake takes place on an upper band, a band gap can decrease in width more abruptly (Hussein, 2009b).

In Fig. 3A, we observe that when the level of damping exceeds a certain value, the lower band gets cut-off in the wavenumber domain, i.e., it does not span the entire first irreducible Brillouin zone (IBZ), $\kappa a \in [0,\pm\pi]$. Moreover, since the upper band eventually disappears we get a wavenumber band gap, i.e., a wavenumber range where waves are prohibited from propagation. This phenomenon is analogous to the well-known concept of a

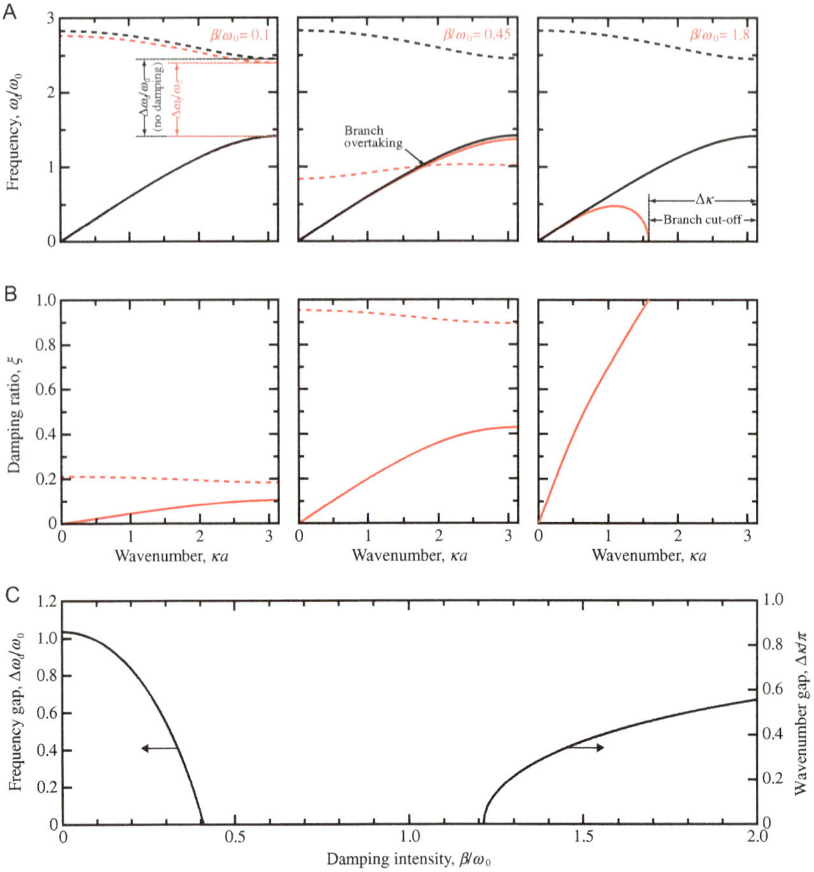

Fig. 3 (A) Dispersion diagram for the phononic crystal shown in Fig. 2 illustrating the dynamic response of the material to various damping levels: undamped, $\beta = 0$ (*black*) and damped, $\beta > 0$ (*red*). In particular, the descent of frequency bands with increased damping is highlighted, ultimately resulting in branch overtaking and band cut-off. (B) The corresponding damping ratio diagram which is unique to complex frequencies and is a measure of how rapidly the wave amplitude decays—no propagation is possible for $\xi(\kappa) \geq 1$. (C) As damping increases, the frequency bands collapse with higher frequencies outpacing lower ones, narrowing and ultimately closing the band gap and initiating band overtaking. Subsequently, beginning with higher frequency bands, a gap in the wavenumber spectrum opens, resulting in band cut-off.

frequency band gap. The damping-induced band-overtaking and band cut-off phenomena clearly present opportunities for design, building on already existing methodologies at the unit-cell level and/or a combination of the unit-cell and structural levels. However, both phenomena require substantial levels of prescribed damping.

2. METADAMPING: DAMPING EMERGENCE

2.1 Locally Resonant Metamaterials

In the previous section, we investigated the dynamics of phononic materials with and without the influence of damping; however, the continuous and discrete examples studied thus far have been of the phononic crystal type. The unit-cell structure of a PC interacts with traveling elastic waves to promote interferences, opening up band gaps. As a result, there is a dependency on unit-cell size to open band gaps.

Over the past two decades, metamaterials, whose extraordinary dynamic and mechanical properties originate from a uniquely tailored internal architecture, have spurred research in the acoustics (Liu et al., 2000), electromagnetics (Pendry, Holden, Robbins, & Stewart, 1999; Smith, Padilla, Vier, Nemat-Nasser, & Schultz, 2000), and, more recently, mechanics (Christensen, Kadic, Kraft, & Wegener, 2015) fields. Among the most intriguing applications is the possibility of a perfect lens (Pendry, 2000). Returning to our interest in the dynamical aspects of phononic materials, much of the remarkable properties of metamaterials, such as negative effective mass/density, elastic/bulk modulus, refractive index, etc.—are often the homogenized manifestation of subwavelength resonances engineered into each unit cell. These resonating bodies act to localize (rather than scatter/reflect) the wave energy to open band gaps and may take the form of heavy, elastically coated spheres in an epoxy matrix (Liu et al., 2000), pillars (Bilal & Hussein, 2013; Pennec, Djafari-Rouhani, Larabi, Vasseur, & Ladky-Hennion, 2008; Wu, Huang, Tsai, & Wu, 2008), voids in an epoxy matrix (Wang, Wen, Wen, Shao, & Liu, 2004), localized oscillators distributed along a rod (Al Ba'ba'a & Nouh, 2017; Khajehtourian & Hussein, 2014), among other possibilities. In the following, we investigate the utility of locally resonant metamaterials in addressing a long-standing materials challenge involving damping.

2.2 Metadamping in Locally Resonant Metamaterials

A variety of applications demand materials with simultaneously high damping (i.e., vibration/shock suppression) and mechanical stiffness (i.e., load-bearing capability); however, for traditional materials, a gain in one is often at the expense of the other. Consider the load-bearing and damping properties of steel and rubber as one of many such material comparisons (see Fig. 4). Following a "best-of-both-worlds" approach, even composites

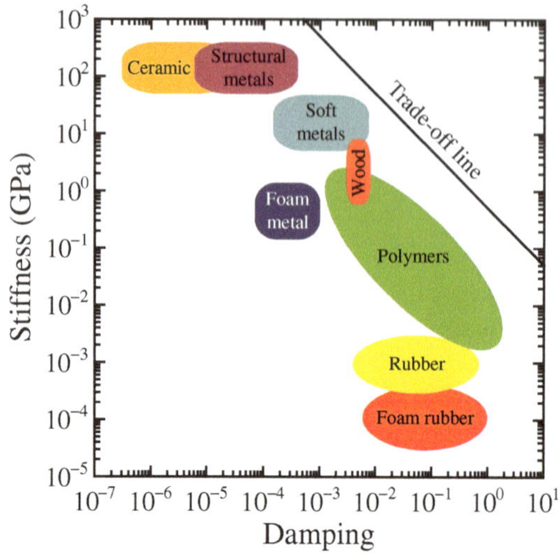

Fig. 4 Ashby chart of stiffness vs damping of common materials illustrating the stiffness-damping trade-off.

composed of stiff and damped constituents are compelled by inherent properties and filling fraction trade-off to sacrifice one property to promote the other.

In addition to their previously mentioned attributes and applications in the absence of material damping, metamaterials may exhibit a greater capacity for wave attenuation than traditional, nonarchitected materials and PCs when damping is taken into account. This gives metamaterials the potential to enable very high levels of dissipation with sustained load-bearing capacity. This emergent damping capacity, termed *metadamping* (Hussein & Frazier, 2013b), applies across a broad spectrum rather than a narrow set of frequencies and has been demonstrated in metamaterials supporting monopolar (Frazier & Hussein, 2015) and dipolar (Frazier & Hussein, 2015; Hussein & Frazier, 2013b) resonances, and resonances in more complex configurations (Antoniadis, Chronopoulos, Spitas, & Koulocheris, 2015; DePauw, Al Ba'ba'a, & Nouh, 2018). In the following, we illustrate the metadamping phenomenon, beginning with metamaterials based on local resonance and then through other architectures.

Fig. 5 shows the initial three material architectures to be considered. Fig. 5A is a representative phononic crystal which relies exclusively on scattering and interference between forward and backward propagating waves for its dispersive properties, notably, the band gap. Fig. 5B and C are two prototypical metamaterial designs whose dynamics are affected by local

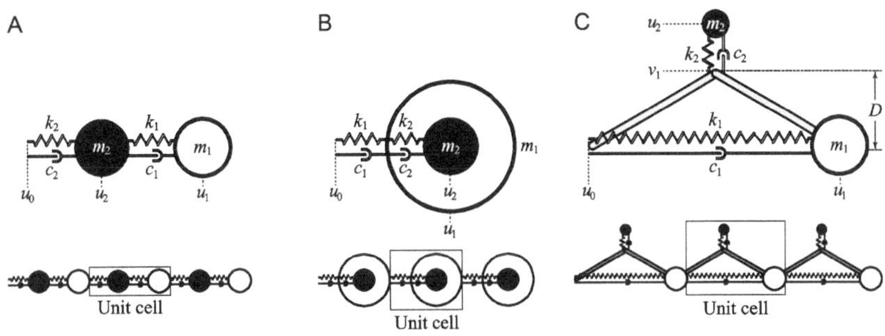

Fig. 5 Discrete models of (A) a phononic crystal (utilizing scattering and wave interferences), and (B) dipolar and (C) monopolar metamaterials (utilizing local resonance). These lumped-parameter models are simple, yet retain the essential physics of their laboratory counterparts.

resonances, particularly dipolar and monopolar, in addition to wave interference due to the periodicity. In acoustics, a monopole radiates vibrational waves equally well in all directions; thus, the truss structure in the monopole metamaterial design in Fig. 5C, which equally applies the influence of the resonator to both ends of the unit cell. A dipole consists of two monopoles of equal magnitude but opposite phase and separated by a subwavelength distance so that vibrational waves of opposite phase radiate well in two directions. This is modeled by the metamaterial unit cell in Fig. 5B. The reader is referred to the article by Russell, Titlow, and Bemmen (1999) for more information on monopoles and dipoles.

For each metamaterial model, the equation of motion in Eq. (2) is structurally the same; however, consistent with the unique interaction among the degrees of freedom of each model, the damping and stiffness matrices are defined differently from the phononic crystal matrices introduced earlier (the mass matrix is the same). For the metamaterial with dipolar resonance, the damping and stiffness matrices become

$$\mathbf{C} = c_2 \begin{bmatrix} 1/r_c + 1 & -1 & -1/r_c \\ -1 & 1 & 0 \\ -1/r_c & 0 & 1/r_c \end{bmatrix} \tag{11a}$$

$$\mathbf{K} = k_2 \begin{bmatrix} 1/r_k + 1 & -1 & -1/r_k \\ -1 & 1 & 0 \\ -1/r_k & 0 & 1/r_k \end{bmatrix} \tag{11b}$$

For the metamaterial with monopolar resonance, they are defined as

$$\mathbf{C}=c_2\begin{bmatrix} 1/r_c+\delta/16 & \sqrt{\delta}/4 & -(1/r_c+\delta/16) \\ \sqrt{\delta}/4 & 1 & -\sqrt{\delta}/4 \\ -(1/r_c+\delta/16) & -\sqrt{\delta}/4 & 1/r_c+\delta/16 \end{bmatrix} \tag{12a}$$

$$\mathbf{K}=k_2\begin{bmatrix} 1/r_k+\delta/16 & \sqrt{\delta}/4 & -(1/r_k+\delta/16) \\ \sqrt{\delta}/4 & 1 & -\sqrt{\delta}/4 \\ -(1/r_k+\delta/16) & -\sqrt{\delta}/4 & 1/r_k+\delta/16 \end{bmatrix} \tag{12b}$$

where $\delta = (a/D)^2$ is an additional control parameter for the monopolar resonance (a denotes the lattice spacing).

With the above construction, the three models share the same transformation matrix $\mathbf{T}$ as the phononic crystal, which, when applied to the equation of motion, Eq. (2), gives its reduced form in terms of the essential degrees of freedom, Eq. (5).

To facilitate a proper comparison of material performance, notice that each model contains the same damping elements and number of degrees of freedom, i.e., an advantage in the form of more effective damping mechanisms and/or more interactions is not held by either model. In addition, the three models are statically equivalent, that is, they are of equal mass and effective static stiffness, $k_{\text{eff}} = c_0^2(m_1 + m_2)/a$, where c_0 is extracted from the linear portion of the undamped dispersion diagram where $c_0 \approx \omega/\kappa$ (see Section 4 for an in-depth analysis of static equivalence). This requires $k_{1,\ 2}$ to differ between the models, however, r_k is held constant. Specifically, for the dipolar metamaterial, $\omega_0 = 46.19$ rad/s; for the monopolar metamaterial with $\delta = 1$, $\omega_0 = 23.09$ rad/s. For the present set of material parameters, $c_0 = 40$.

In Fig. 6, we compare the response of each model to a common damping intensity, $\beta/\omega_0 = 0.1$. The results show that there are shifts in the frequency band diagrams of all three models due to damping which are consistent with what was seen in Fig. 3A—greater shifts at higher frequencies corresponding to higher damping ratios. However, despite the static equivalence and equally prescribed viscous damping, we observe that, compared to the representative PC, the metamaterials exhibit higher damping ratio values (i.e., higher dissipation) across much of the wavenumber domain. This is an indication of a considerable amplification, or emergence, of dissipation in locally resonant metamaterials compared to their phononic crystal counterparts and

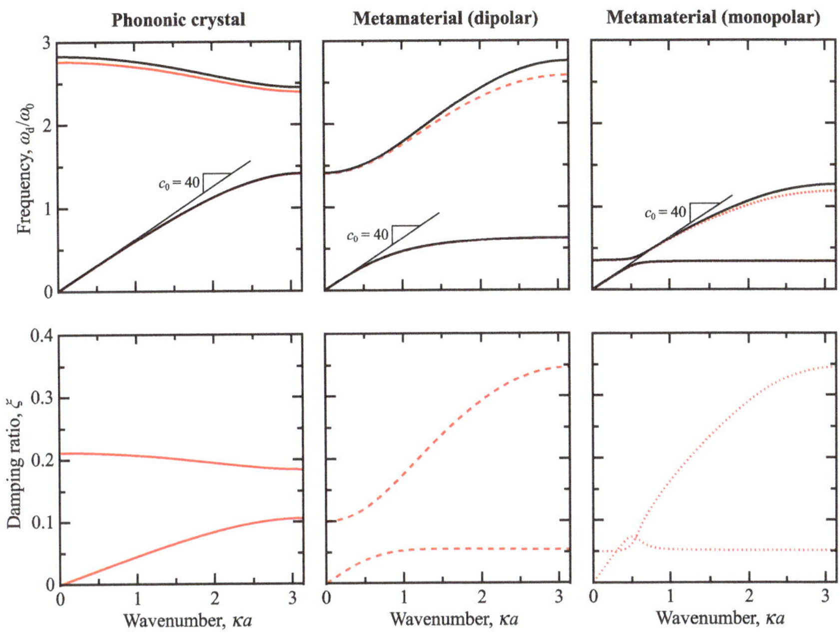

Fig. 6 Frequency and damping ratio dispersion of statically equivalent [equal mass and long-wavelength stiffness (equivalent speed of sound)] phononic crystal and metamaterial models.

showcases the effect of the internal structure design. For a more comprehensive indicator of the damping capacity of each material model, we determine the average damping ratio over the whole of the wavenumber domain and across all modes:

$$\xi_{\text{avg}} = \frac{1}{2\pi}\int_0^{\pi} [\xi_1(\kappa a) + \xi_2(\kappa a)]\text{d}\kappa a \tag{13}$$

The results—$\xi_{\text{avg}}^{\text{PC}} = 0.13$ for the phononic crystal, $\xi_{\text{avg}}^{\text{DP}} = 0.20$ for the dipolar metamaterial, and $\xi_{\text{avg}}^{\text{MP}} = 0.39$ for the monopolar metamaterial—reveal the superior dissipation performance of metamaterials, particularly those possessing monopolar resonances. This enhanced dissipation in metamaterials defines the metadamping phenomenon.

Fig. 7 illustrates the robustness of the metadamping phenomenon by plotting ξ_{avg} for various c_0 values. In addition, the impact of the parameter δ unique to the monopolar model is demonstrated. For the case $c_0 = 40$, even as $\delta \to 0$, monopolar metadamping still outperforms the dipolar case.

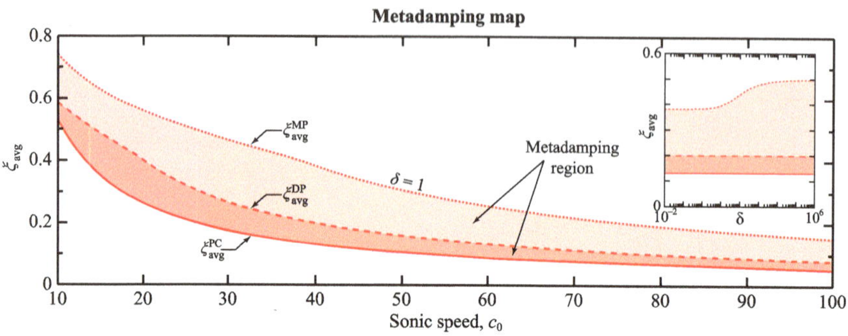

Fig. 7 Metadamping map characterizing the sonic speed dependency of the emergent increase in damping capacity in resonance-based metamaterials.

In this section, we have demonstrated the concept of metadamping theoretically due to the presence of local resonance in the internal structure of the phononic material. This finding has far reaching implications on the design of materials for numerous applications that require the reduction, mitigation, or absorption of vibrations, shock, and/or sound. While the analysis has been presented in the context of simple mass-spring-dashpot periodic chains, it can be readily extended to practical realizations of locally resonant acoustic metamaterials. Examples include material structures that utilize: heavy inclusions with compliant coatings, soft inclusions, split-resonators, inertial amplifiers, pillars, holey cylinders, and suspended masses. Other concepts for enhancing damping while retaining stiffness may be applied in conjunction with the inclusion of local resonators leading to an additive effect. While this discussion has presented metadamping as a desirable effect, an awareness and understanding of the phenomenon will aid in mitigation where it is not advantageous. Finally, while the concept of metadamping has been presented in the context of a mechanical problem, in principle it is also applicable to other disciplines in materials physics that involve both resonance and dissipation.

3. UNCONVENTIONAL METAMATERIAL DESIGNS

In pursuit of novel functionalities, unconventional metamaterial configurations have recently emerged that exhibit broader tunability of wave dispersion features and, at times, offer a pathway for physical realization of locally resonant mechanisms (Chen, Hu, & Huang, 2017; Liu, Chan, & Sheng, 2005). The overarching theme of these novel metamaterials is the combination of phononic features and intrinsic resonators placed in unique and unprecedented configurations. Among these is the notion of resonator-to-resonator interaction with examples that range from spring-mass metamaterials with discretely coupled resonators (Hu, Tang, Das, Gao, & Liu, 2017) to elastic metamaterial beams with interconnected resonators (Beli, Arruda, & Ruzzene, 2018). In addition, other configurations feature the presence of nonlocal resonances which, rather than being confined inside an outer mass, directly interact with the preceding and following masses in a periodic chain (DePauw et al., 2018).

3.1 Nonlocal Metamaterial Design

In this section, we investigate a new class of metamaterials which comprises a nonlocal resonator and combines hybrid attributes from dipolar metamaterials (DP) and PC (DePauw et al., 2018). The new configuration—henceforth referred to as *nonlocal metamaterial* (NL)—consists of two masses m_1 and m_2. Each two consecutive masses m_1 are connected via a spring k_1, while each mass m_2 is sandwiched between two neighboring masses m_1 and connected to each of them via a spring k_2, as depicted in Fig. 8. It is noted that the PC and DP configurations may be realized from the NL metamaterial configuration by setting k_1 or one of the k_2 springs equal to zero, respectively. The dashpots c_1 and c_2 are added in parallel to the respective springs in the damped case to characterize the energy dissipation in the NL metamaterial, which is capable (as will be shown later) of exercising considerably greater metadamping effects than both the PC and the DP cases with the suitable choice of parameters.

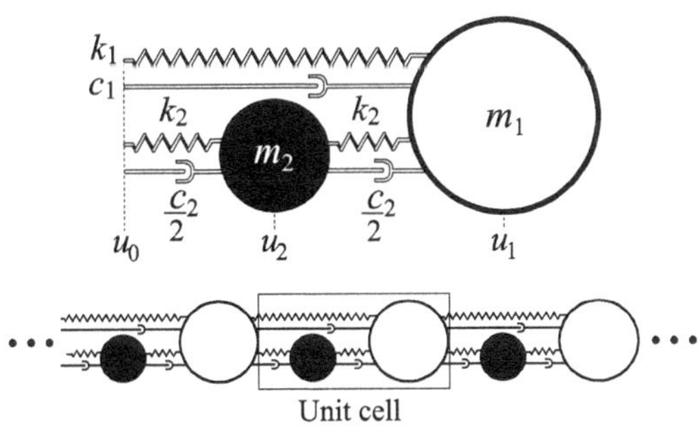

Fig. 8 Discrete model of a nonlocal metamaterial (NL) cell which repeats infinitely along one dimension.

3.2 Dispersion Characteristics and Transitions

We start by highlighting the dispersion characteristics of the undamped NL metamaterial under both free wave and driven wave conditions. In the most general form, the undamped dispersion relations are given by

$$\alpha_2\omega^4 - \alpha_1\omega^2 + \alpha_0 = 0 \tag{14}$$

where $\alpha_2 = m_1m_2$ and the coefficients α_1 and α_0 for the different phononic materials are listed in Table 1.

The general dispersion relation provided in Eq. (14) depicts the free wave formulation and, for the NL case, reads

$$\omega^4 - \left(2(1+r_m) + \frac{r_m}{r_k}\alpha\right)\omega_0^2\omega^2 + \omega_0^4 r_m(1+2/r_k)\alpha = 0 \tag{15}$$

where the acoustical ω_- and optical ω_+ branches are given in closed form by

$$\omega_\pm^2 = \frac{\omega_0^2}{2}\left[\left(2(1+r_m) + \frac{r_m}{r_k}\alpha\right) \pm \sqrt{\left(2(1+r_m) + \frac{r_m}{r_k}\alpha\right)^2 - 4r_m(1+2/r_k)\alpha}\right] \tag{16}$$

For the free wave formulation, the dispersion curves are obtained by computing the roots of Eq. (16) over a range of $\kappa a \in [0,\pi]$ corresponding to the IBZ. Upon inspecting the results, it can be observed that the optical dispersion branch can be manipulated to exhibit a negative or positive group velocity depending on the choice of parameters, thus resembling either a PC or a resonant material, respectively. Such a feature implies the presence of a turning point where the optical branch becomes flat (i.e., $\frac{\partial\omega}{\partial\kappa} = 0$); a phenomenon that has been observed in systems with parametric amplification (Cassedy, 1967). A similar behavior has also been captured in inertially amplified locally resonant metamaterials (Al Ba'ba'a, DePauw, Singh, & Nouh, 2018). The turning point can be analytically extracted by using

Table 1 Expressions for α_0 and α_1 for Different Phononic Material ($\alpha = 2(1 - \cos\kappa a)$)

	PC	DP	MP	NL
α_0	$k_1k_2\alpha$	$k_1k_2\alpha$	$k_1k_2\alpha$	$k_2(2k_1 + k_2)\alpha$
α_1	$(k_1+k_2)(m_1+m_2)$	$k_2(m_1+m_2) + k_1m_2\alpha$	$m_1k_2 + m_2(k_1 + \frac{\delta}{16}k_2)\alpha$	$2k_2(m_1+m_2) + k_1m_2\alpha$

the derivative of the dispersion relation in Eq. (15) with respect to κ (Liu & Hussein, 2012)

$$4\omega^3\frac{\partial\omega}{\partial\kappa}-\omega_0^2\left[2\left(2(1+r_m)+\frac{r_m}{r_k}\alpha\right)\omega\frac{\partial\omega}{\partial\kappa}+2a\left(\frac{r_m}{r_k}\sin\kappa a\right)\omega^2\right]$$
$$+2a\omega_0^4r_m(1+2/r_k)\sin\kappa a=0 \quad (17)$$

Given that a flat branch dictates that $\frac{\partial\omega}{\partial\kappa}=0$, Eq. (17) simplifies to

$$\left(\omega_0^2(1+2/r_k)-\omega^2/r_k\right)\sin\kappa a=0 \quad (18)$$

which indicates that a zero group velocity exists when $\sin\kappa a=0$ at $\kappa a=\ell\pi$, where $\ell=0,1,2,\ldots$, as well as for the following condition

$$\omega=\omega_0\sqrt{2+r_k} \quad (19)$$

Substituting Eq. (19) into Eq. (15), it can be shown that the relationship between the stiffness and mass ratios which governs the turning point of the optical branch is given by

$$r_k=2r_m \quad (20)$$

Alternatively, the same condition associated with the turning point can be found by equating the roots of the optical branch ω_+ in Eq. (16) at $\kappa a=0$ and $\kappa a=\pi$ (i.e., $\alpha=0$ and $\alpha=4$, respectively); which results in

$$\omega_+=\omega_0\sqrt{2(1+r_m)} \quad (21a)$$

$$\omega_+=\omega_0\sqrt{2r_m(1+2/r_k)} \quad (21b)$$

By equating Eqs. (21a) and (21b), the same condition to that given by Eq. (20) is found. If $2r_m > r_k$, the NL optical branch behaves in a manner consistent with resonant metamaterials while for $2r_m < r_k$, it resembles the optical branch of a PC. As a consequence, the lower and upper bounds of the resultant band gap, ω_l and ω_u, depend on whether the NL metamaterial mimics a PC or a resonant metamaterial. While ω_l can always be extracted from the dispersion solution at $\kappa a=\pi$, the value of ω_u will obviously depend on the shape of the optical branch and can correspond to the solution at either $\kappa a=0$ or $\kappa a=\pi$. The following two equations summarize all the possible scenarios and provide comprehensive expressions for ω_l and ω_u

$$\omega_l=\min\left(\sqrt{2r_m(1+2/r_k)},\sqrt{2}\right)\omega_0 \quad (22a)$$

$$\omega_u=\min\left(\sqrt{2(1+r_m)},\max\left(\sqrt{2r_m(1+2/r_k)},\sqrt{2}\right)\right)\omega_0 \quad (22b)$$

It is critical to note that when $\alpha = 4$ in Eq. (16), the acoustical branch solution can be shown to be $\omega_- = \sqrt{2}\omega_0$. It is also evident from Eqs. (22a) and (22b) that when an NL behaves like a PC, it is possible for $\sqrt{2r_m(1+2/r_k)}$ to be either greater or less than $\sqrt{2}$. In the rare occasion where $\sqrt{2r_m(1+2/r_k)} < \sqrt{2}$, this dispersion solution appears in the acoustical branch and is otherwise located in the optical branch. In cases where the upper limit $\omega_u = \sqrt{2}\omega_0$, this requires that $r_m(1 + 2/r_k) < 1$, which generates the following condition

$$r_m < \frac{r_k}{2 + r_k} \tag{23}$$

which is a subset of the $2r_m < r_k$ case with r_m being less than unity. Interestingly, at the critical point $r_m = \frac{r_k}{2+r_k}$, the band gap of the NL metamaterial vanishes and the dispersion relation has the repeated solutions $\omega = \sqrt{2}\omega_0$ at $\kappa a = \pi$ (i.e., the discriminant of Eq. (16) becomes equal to zero). In such a case, the NL dispersion resembles that of a homogenous monatomic lattice. To summarize the previous discussion, it is concluded that one of the bandgap limits in the NL metamaterial has to be located at $\omega = \sqrt{2}\omega_0$. The different scenarios accompanying the NL dispersion relation discussed here are graphically summarized in Fig. 9.

For completeness, we also investigate the driven wave formulation which generates the wavenumber κa as an output to a prescribed frequency, i.e., $\kappa a(\omega)$. By recasting Eq. (15) in the form: $\kappa a = \cos^{-1}\Phi(\omega)$, the explicit form of the function $\Phi(\omega)$ for NL metamaterials can be written as

$$\Phi^{\mathrm{NL}} = 1 + \frac{r_k\omega^2}{2r_m\omega_0^2}\left(\frac{\omega^2 - 2\omega_0^2(1+r_m)}{\omega_0^2(2+r_k) - \omega^2}\right) \tag{24}$$

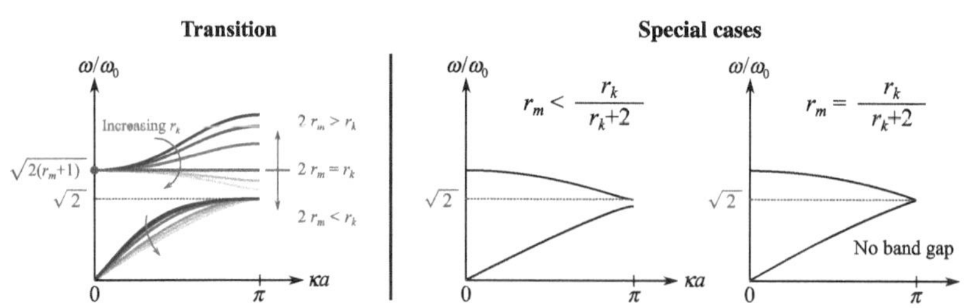

Fig. 9 Dispersion transitions in an NL metamaterial with different mass and stiffness ratios.

By inspecting Eq. (24), the presence of an antiresonance in the system can be determined from the discontinuities in the Φ^{NL} function, specifically the roots of its denominator. The NL metamaterial has an antiresonance frequency ω_R which is given by

$$\omega_R = \omega_0\sqrt{2 + r_k} \tag{25}$$

which, as can be seen, is a function of the stiffness ratio r_k, for a given ω_0. In addition, the driven wave problem facilitates the process of extracting the extreme points of the dispersion from the roots of $\frac{\partial \Phi^{\mathrm{NL}}}{\partial \omega}$

$$\frac{\partial \Phi^{\mathrm{NL}}}{\partial \omega} = \omega\left(\omega^4 - 2\omega_0^2(2 + r_k)\omega^2 + 2\omega_0^4(1 + r_m)(2 + r_k)\right) = 0 \tag{26}$$

where nonzero solutions for ω in Eq. (26) are found as

$$\hat{\omega}_\pm = \omega_0\sqrt{(2 + r_k) \pm \sqrt{(2 + r_k)(r_k - 2r_m)}} \tag{27}$$

It is worth noting that $\omega_R = \omega_0\sqrt{2 + \kappa}$ is a point of maximum attenuation which corresponds to the discriminant of Eq. (27) being zero in the case where $2r_m = r_k$. The structure of Φ^{NL} ensures that the value of $\Phi^{\mathrm{NL}}(\hat{\omega}_\pm) > 1$, which in turn provides information regarding extrema of the imaginary component of the wavenumber $\kappa_{\mathrm{I}} a$, leading to points where the gradient is zero. The analysis presented here for driven waves can be similarly extended to the different types of phononic materials. For convenience, expressions for $\Phi(\omega)$, ω_R, and $\hat{\omega}$ are given in Table 2.

A comparison between the PC, DP, and NL systems is presented in Fig. 10. The four rows in the figure represent the different dispersion

Table 2 Expressions for Φ for the Phononic Crystal and DP and MP Metamaterials

	$\Phi(\omega)$	ω_R	$\hat{\omega}$
PC	$1 + \frac{r_k\omega^2}{2r_m\omega_0^2}\left(\frac{\omega^2}{\omega_0^2} - (1 + r_m)(1 + 1/r_k)\right)$	—	$\omega_0\sqrt{\frac{1}{2}(1 + r_m)(1 + 1/r_k)}$
DP	$1 + \frac{r_k\omega^2}{2r_m\omega_0^2}\left(\frac{\omega^2 - \omega_0^2(1 + r_m)}{\omega_0^2 - \omega^2}\right)$	ω_0	No valid solution
MP	$1 + \frac{r_k\omega^2}{2r_m\omega_0^2}\left(\frac{(\omega^2 - \omega_0^2)}{\omega_0^2 - (1 + \frac{\delta}{16}r_k)\omega^2}\right)$	$4\omega_0\sqrt{\frac{1}{16 + \delta r_k}}$	No valid solution

transitions that are unique to the NL design. The mass and stiffness ratio combinations $[r_m, r_k]$ are $[3, 1]$, $[\frac{3}{2}, 3]$, $[1, 3]$, and $[\frac{1}{3}, 1]$ for Fig. 10A–D, respectively. In each case, the corresponding dispersion diagrams for the PC and DP systems are shown for comparison. In the latter, while the location of the branches (as well as the local resonance ω_R) changes from one row to the next, the overall dispersion shapes are preserved. With a focus on the rightmost column corresponding to the NL metamaterial, and starting with the $2r_m > r_k$ case (Fig. 10A), the dispersion behavior of the NL resembles that of a conventional DP metamaterial and the resonance $\omega_R = \omega_0\sqrt{2+\kappa}$ is located inside the band gap. At the turning point (Fig. 10B), $\omega_R = \omega_0\sqrt{2+\kappa}$ coincides with the upper limit of the band gap; canceling out the optical branch as a consequence. This also reaffirms the result obtained from Eq. (19) when the group velocity $\frac{\partial\omega}{\partial\kappa} = 0$. By revisiting Eq. (24) with $2r_m = r_k$, the numerator and denominator of the far right term match and the equation reduces to $\Phi^{NL} = 1 - \frac{\omega^2}{\omega_0^2}$ which explains the lack of an ω_R in such case. Furthermore, a confined band gap ceases

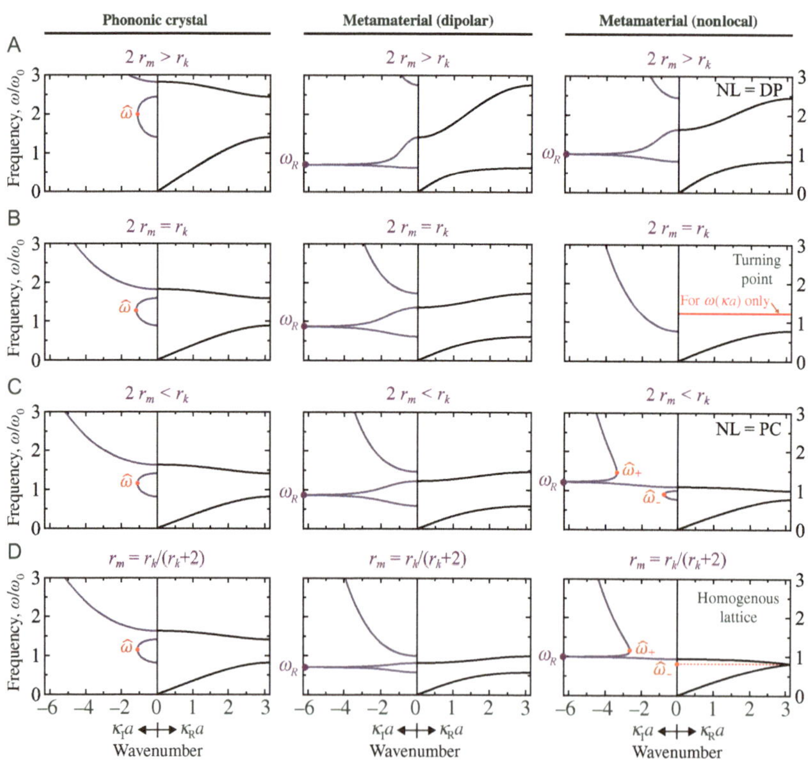

Fig. 10 Dispersion relations for PC (*left*), DP (*middle*), and NL (*right*) for different mass and stiffness ratios: (A) $r_m = 3$, $r_k = 1$; (B) $r_m = \frac{3}{2}$, $r_k = 3$; (C) $r_m = 1$, $r_k = 3$; and (D) $r_m = \frac{1}{3}$, $r_k = 1$.

to exist. Instead, an unbounded stop band starts immediately at $\omega = \sqrt{2}\omega_0$. For the case when $2r_m < r_k$, $\hat{\omega}_-$ represents the Bragg band-gap maximum attenuation frequency while $\hat{\omega}_+$ reflects an inflection point in the unbounded stop band region, as captured in Fig. 10C. In the final scenario, at $r_m = \frac{r_k}{r_k + 2}$, $\hat{\omega}_-$ coincides with $\omega = \sqrt{2}\omega_0$ when the NL behaves akin to a homogenous monatomic lattice, thus closing the band gap, whereas the inflection point exists at $\hat{\omega}_+ = \omega_0\sqrt{2(1 + r_k)}$ as can be seen in Fig. 10D.

4. STATIC EQUIVALENCE AT THE LONG-WAVE SPEED LIMIT

To enable a fair comparison between the dissipation performance of different phononic materials, the systems should be statically equivalent, i.e., exhibit an identical long-wave (sonic) speed c_0. This condition has been imposed in Section 2.2. In this section, we provide a formal derivation of static equivalence. We start by analytically deriving expressions for c_0 corresponding to the different phononic materials considered, from their respective dispersion relations.

The two solutions for Eq. (14) are given by

$$\omega_\pm = \sqrt{\frac{\alpha_1 \pm \sqrt{\alpha_1^2 - 4\alpha_2\alpha_0}}{2\alpha_2}} \tag{28}$$

Of interest here is the solution corresponding to the acoustical branch ω_-, which can be rewritten as

$$\omega_- = \sqrt{\frac{1 - \sqrt{1 - \frac{4\alpha_2\alpha_0}{\alpha_1^2}}}{2\frac{\alpha_2}{\alpha_1}}} \tag{29}$$

To find c_0, the ratio $\frac{\omega}{\kappa}$ is obtained near $\kappa a = 0$. As can be seen in Table 1, α_0 is a function of the wavenumber κ and its value approaches zero as $\kappa a \to 0$, resulting in $\frac{4\alpha_2\alpha_0}{\alpha_1^2} \ll 1$. It is important to note that although α_1 may also be a function of κ, depending on the type of phononic material considered, it always has a constant nonzero value when $\alpha = 0$. As a result, the assumption $\frac{4\alpha_2\alpha_0}{\alpha_1^2} \ll 1$ is not affected. Using the binomial approximation $\sqrt{1 - \frac{4\alpha_2\alpha_0}{\alpha_1^2}} \approx 1 - \frac{1}{2}\left(\frac{4\alpha_2\alpha_0}{\alpha_1^2}\right)$, Eq. (29) may be simplified to

$$\omega = \sqrt{\frac{\alpha_0}{\alpha_1}} \tag{30}$$

Further, by utilizing the identity $1 - \cos\kappa a = 2\sin^2\frac{\kappa a}{2}$, and for small values of κa (i.e., $\sin^2\frac{\kappa a}{2} = \left(\frac{\kappa a}{2}\right)^2$), obtaining an expression for c_0 for the PC case is straight forward and takes the form

$$\omega = \kappa a\sqrt{\frac{k_1 k_2}{(m_1 + m_2)(k_1 + k_2)}} \rightarrow c_0^{\mathrm{PC}} = a\omega_0^{\mathrm{PC}}\sqrt{\frac{r_m}{(1 + r_m)(1 + r_k)}} \tag{31}$$

For the remaining types of phononic materials, a Taylor series approximation around $\kappa a = 0$ is needed to further reduce Eq. (30) since both α_0 and α_1 are functions of κa. Now, let us rewrite α_0 and α_1 as $\alpha_0 = \bar{\alpha}_0\alpha$ and $\alpha_1 = \bar{\alpha}_1\alpha + \bar{\bar{\alpha}}_1$, respectively, where the corresponding values of the variables $\bar{\alpha}_0$, $\bar{\alpha}_1$, and $\bar{\bar{\alpha}}_1$ can be extracted from Table 1 by matching the coefficients. Substituting $\alpha = 4\sin^2\frac{\kappa a}{2}$, Eq. (30) now reads

$$\omega = \sqrt{\frac{4\bar{\alpha}_0\sin^2\dfrac{\kappa a}{2}}{\bar{\bar{\alpha}}_1 + 4\bar{\alpha}_1\sin^2\dfrac{\kappa a}{2}}} = \sqrt{f(\kappa a)} \tag{32}$$

The term under the square root in Eq. (32) can be expanded using a Taylor series around $\kappa a = 0$, which yields

$$f(\kappa a) \simeq f(0) + f'(0)\kappa a + \frac{1}{2}f''(0)(\kappa a)^2 \tag{33}$$

where

$$f'(\kappa a) = \frac{2\bar{\bar{\alpha}}_1\bar{\alpha}_0\sin\kappa a}{\left(\bar{\bar{\alpha}}_1 + 4\bar{\alpha}_1\sin^2\dfrac{\kappa a}{2}\right)^2} \tag{34a}$$

$$f''(\kappa a) = 2\bar{\alpha}_0\bar{\bar{\alpha}}_1\frac{\bar{\alpha}_1(2\cos\kappa a + \cos 2\kappa a - 3) + \bar{\bar{\alpha}}_1\cos\kappa a}{\left(4\bar{\alpha}_1\sin^2\left(\dfrac{\kappa a}{2}\right) + \bar{\bar{\alpha}}_1\right)^3} \tag{34b}$$

Upon evaluating the first and second derivatives of $f(\kappa a)$ at $\kappa a = 0$, it can be shown that

$$\omega = \kappa a\sqrt{\frac{\bar{\alpha}_0}{\bar{\bar{\alpha}}_1}} \tag{35}$$

and the explicit formulae for the long-wave speed c_0 of DP, MP, and NL metamaterials can be given by

$$c_0^{\mathrm{DP}} = a\omega_0^{\mathrm{DP}}\sqrt{\frac{r_m}{r_k(1 + r_m)}} \tag{36a}$$

$$c_0^{\mathrm{MP}} = a\omega_0^{\mathrm{MP}}\sqrt{\frac{r_m}{r_k}} \tag{36b}$$

$$c_0^{\mathrm{NL}} = a\omega_0^{\mathrm{NL}}\sqrt{\frac{(2/r_k+1)}{2(1+1/r_m)}} \tag{36c}$$

The above equations may be rearranged with the long-wave speed c_0 as an input to obtain the corresponding ω_0, resulting in

$$\omega_0^{\mathrm{PC}} = \frac{c_0^{\mathrm{PC}}}{a}\sqrt{(1+r_k)(1+1/r_m)} \tag{37a}$$

$$\omega_0^{\mathrm{DP}} = \frac{c_0^{\mathrm{DP}}}{a}\sqrt{r_k(1+1/r_m)} \tag{37b}$$

$$\omega_0^{\mathrm{MP}} = \frac{c_0^{\mathrm{MP}}}{a}\sqrt{\frac{r_k}{r_m}} \tag{37c}$$

$$\omega_0^{\mathrm{NL}} = \frac{c_0^{\mathrm{NL}}}{a}\sqrt{\frac{2(1+1/r_m)}{(2/r_k+1)}} \tag{37d}$$

By setting c_0 of the PC as the benchmark for comparison, and maintaining identical c_0 values for the rest of the designs, alternative expressions for the frequency ω_0 as a function of ω_0^{PC} for the metamaterial counterparts can be derived as

$$\omega_0^{\mathrm{DP}} = \omega_0^{\mathrm{PC}}\sqrt{\frac{r_k}{1+r_k}} \tag{38a}$$

$$\omega_0^{\mathrm{MP}} = \omega_0^{\mathrm{PC}}\sqrt{\frac{r_k}{(1+r_m)(1+r_k)}} \tag{38b}$$

$$\omega_0^{\mathrm{NL}} = \omega_0^{\mathrm{PC}}\sqrt{\frac{2r_k}{(1+r_k)(2+r_k)}} \tag{38c}$$

For instance, if we look specifically at $c_0 = 40$ (see Fig. 6), using Eq. (31) and the parameters $r_m = 3$, $r_k = 1$, and $a = 1$, results in $\omega_0^{\mathrm{PC}} = 65.32$. Finally, by making use of Eqs. (38a)–(38c), the equivalent frequencies ω_0 for each of the DP, MP, and NL are $\omega_0^{\mathrm{DP}} = 46.19$, $\omega_0^{\mathrm{MP}} = 23.09$, and $\omega_0^{\mathrm{NL}} = 37.71$, respectively.

5. METADAMPING IN NONLOCAL METAMATERIALS

In Section 3, the intriguing features associated with the dispersion characteristics of undamped NL metamaterials were thoroughly discussed. Now we examine and highlight the dissipative performance of the NL metamaterial in the presence of damping elements, and draw comparisons with the PC and DP chains (a comparison with the MP system is omitted for brevity). In the presence of the dashpots c_1 and c_2, the mass, damping and stiffness matrices are given by

$$\mathbf{M} = m_2 \begin{bmatrix} 1/r_m & 0 & 0 \\ 0 & 1 & 0 \\ 0 & 0 & 0 \end{bmatrix} \tag{39a}$$

$$\mathbf{C} = c_2 \begin{bmatrix} (1/r_c + 1/2) & -1/2 & -1/r_c \\ -1/2 & 1 & -1/2 \\ -1/r_c & -1/2 & (1/r_c + 1/2) \end{bmatrix} \tag{39b}$$

$$\mathbf{K} = k_2 \begin{bmatrix} (1/r_k + 1) & -1 & -1/r_k \\ -1 & 2 & -1 \\ -1/r_k & -1 & (1/r_k + 1) \end{bmatrix} \tag{39c}$$

The dispersion characteristic of the dissipative NL metamaterial is obtained by invoking the Bloch boundary condition via the transformation matrix $\mathbf{T}$ as introduced earlier, and using the state-space formulation to obtain the complex frequencies λ. Similar to the example presented in Fig. 6, all the systems undergoing comparisons are statically equivalent with a long-wave speed of $c_0 = 40$ and identical stiffness and mass ratios. Recall that the damping intensity ratio is $\beta/\omega_0^{\mathrm{PC}} = 0.1$ from which the damping coefficients are computed as $c_2 = \beta m_2$ and $c_1 = c_2/r_c$. Since the NL metamaterial comprises an extra c_2 damper in each unit cell when compared to the PC and DP configurations, the value of c_2 is halved to maintain a fair comparison, which is reflected in the $\mathbf{C}$ matrix in Eq. (39b). While keeping the same damping coefficients, we investigate the metadamping effects with a second set of parameters leaving r_m unchanged and setting $r_k = 3$. The adjusted ω_0 values for the second set of parameters are $\omega_0^{\mathrm{PC}} = 92.38$, $\omega_0^{\mathrm{DP}} = 80$, and $\omega_0^{\mathrm{NL}} = 50.6$.

Figs. 11 and 12 depict the dispersion curves and the damping ratios across the optical and acoustical bands for the first and second set of parameters, respectively. It is evident from both figures that the overall drop of the optical branch in response to increased damping is larger in the NL than it is in the DP system. Furthermore, the change in the stiffness ratio generates a larger increase in the damping ratio in the NL than in the DP. In both cases, the damping ratios of the NL is either slightly or considerably higher than that of the DP metamaterial, while both metamaterial designs exhibit larger

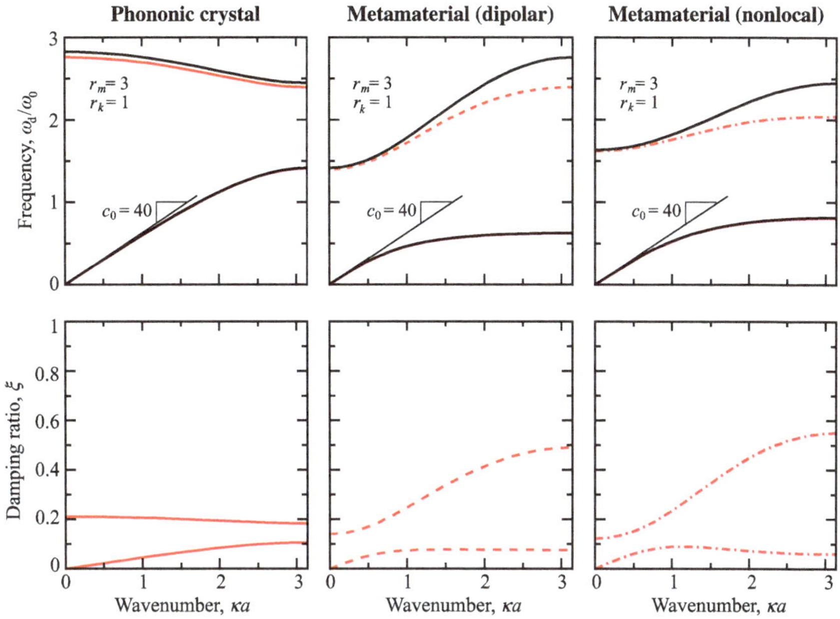

Fig. 11 Dispersion curves and corresponding damping ratios for the damped PC, DP, and NL systems with $r_m = 3$, $r_k = 1$, and $c_0 = 40$.

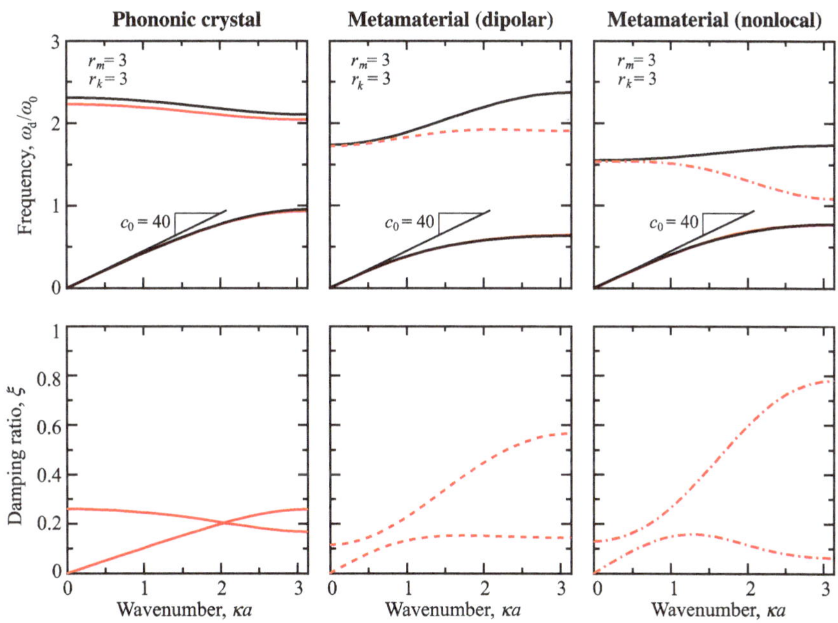

Fig. 12 Dispersion curves and corresponding damping ratios for the damped PC, DP, and NL systems with $r_m = 3$, $r_k = 3$, and $c_0 = 40$.

damping ratios than the PC. Following the earlier analysis, the averaged damping ratio ξ_{avg} is computed from the integration of $(\xi_1 + \xi_2)$ over the wavenumber κa ranging from 0 to π (as can be seen in Eq. (13)). The sonic speed c_0 is swept over a range of values while keeping the same mass and stiffness ratios as well as the damping amount, by adjusting the value of ω_0 for each system, respectively. The outcome of this process is graphically presented in Figs. 13 and 14 which correspond to the cases discussed in Figs. 11 and 12, respectively. It is evident that the NL outperforms the

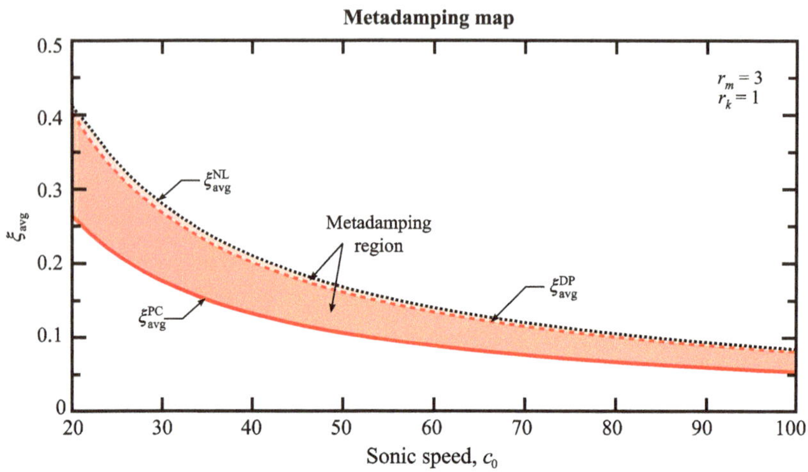

Fig. 13 Metadamping regions in the NL and DP metamaterials (compared to the PC as a datum) for $r_m = 3$, $r_k = 1$.

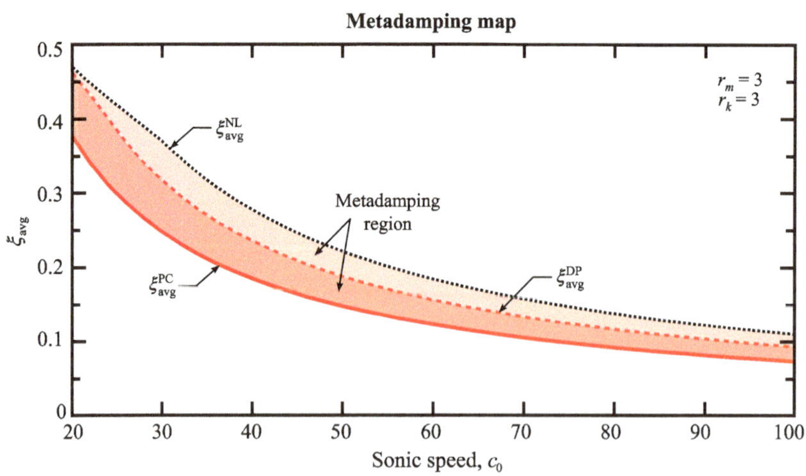

Fig. 14 Metadamping regions in the NL and DP metamaterials (compared to the PC as a datum) for $r_m = 3$, $r_k = 3$.

PC and the DP systems in terms of total metadamping in both cases. The amplified metadamping emergence in the NL is particularly observed with the second set of parameters, validating the damping ratio diagrams presented in Fig. 12.

6. PARAMETRIC STUDY OF METADAMPING

As revealed in the previous section, the change in the stiffness ratio r_k enhanced the metadamping emergence in the NL metamaterial for a prescribed set of damping coefficients and ratio r_c. As such, it is imperative to investigate this phenomenon further across a broader spectrum of properties and material variations, while maintaining to the extent possible reasonable constraints that render such comparisons just. We start by examining the effect of increasing the damping amount (i.e., varying the values of c_1 and c_2) as well as the choice of the ratio r_c (reflecting the damping distribution across the two materials). Fig. 15 combines the effect of the increasing the damping intensity ratio β/ω_0 with different choices of $c_{1,2}$ for the three phononic materials under consideration: the PC, DP, and the NL. Other parameters used in the comparison are identical to those selected in Fig. 12: $r_k = r_m = 3$ and $c_0 = 40$. In each subplot of Fig. 15, three separate cases are considered: (1) $c_2 = 0$ (i.e., $r_c = 0$) and $c_1 = \beta m_2$, (2) $c_1 = 0$ (i.e., $r_c = \infty$) and $c_2 = \beta m_2$, and (3) $r_c = 1$ (with $c_1 = c_2 = \beta m_2/2$). It is noted that in the last case, the damping coefficients are reduced to half their previous values to keep the total damping amount constant across the board. It is first observed that the average damping ratio ξ_{avg} increases linearly with an increase in the damping intensity ratio, irrespective of the phononic material type, for all combinations of c_1 and c_2. However, it can be seen that the damping distribution within the same phononic material

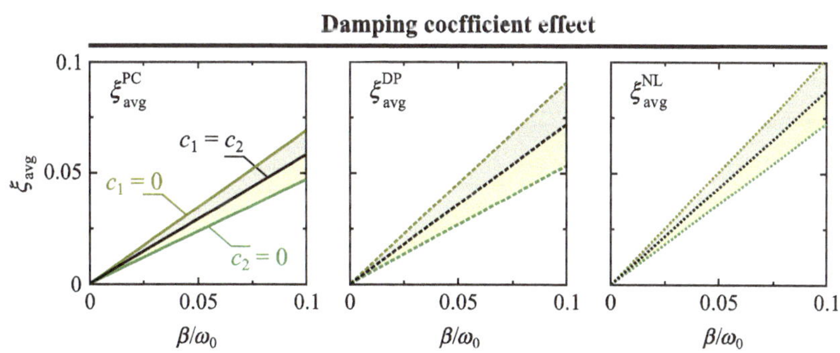

Fig. 15 Metadamping variation with the damping intensity ratio β/ω_0 between the PC, DP, and NL systems for $r_k = r_m = 3$ and $c_0 = 40$.

plays an influential role in the emergent metadamping. Between the two extreme cases corresponding to $r_c = 0$ (i.e., $c_2 = 0$) and $r_c = \infty$ (i.e., $c_1 = 0$), the latter exhibits larger metadamping for the same amount of total damping injected in the system. The metadamping in the case $c_1 = c_2$ lies halfway between the two extreme damping distributions.

With both damping coefficients c_1 and c_2 kept constant, the implications of changing other material properties in the phononic design are also investigated. Fig. 16 shows the variation of metadamping with increasing damping intensities for the two stiffness ratios $r_k = 1$ and $r_k = 3$ examined earlier. As an initial takeaway, it can be seen that the metadamping corresponding to $r_k = 3$ (and reflected by the metadamping metric ξ_{avg}) quantitatively exceeds that of the $r_k = 1$ design for all three systems considered. The previous is also in line with Figs. 11 and 12. In addition, the linear relationship between the damping intensity and the metric ξ_{avg} remains intact. The relative enhancement of the metadamping across the different phononic materials given a prescribed set of parameters may be computed using the metric $\mathbf{Z}$, illustrated in the rightmost panel of Fig. 16. For example,

$$\mathbf{Z}_{\text{DP}}^{\text{NL}} = \xi_{\text{avg}}^{\text{NL}} - \xi_{\text{avg}}^{\text{DP}} \tag{40}$$

defines the relative enhancement (or deterioration) of the emergent metadamping in the NL metamaterial as compared to a corresponding DP metamaterial. Using the newly defined $\mathbf{Z}$ as a metric to assess relative metadamping improvements, or lack thereof, design charts that comprise variations in material properties such as mass, stiffness, and damping ratios, and the interplay between them may be constructed. Investigating the damping distributions across the different constituents of a phononic material allows us to enhance the metadamping outcome. To do so, the stiffness ratio is swept between the values of 0.01–100 and is varied against the ratio r_c. Contours of the relative metadamping metric $\mathbf{Z}$ as a function of both r_k

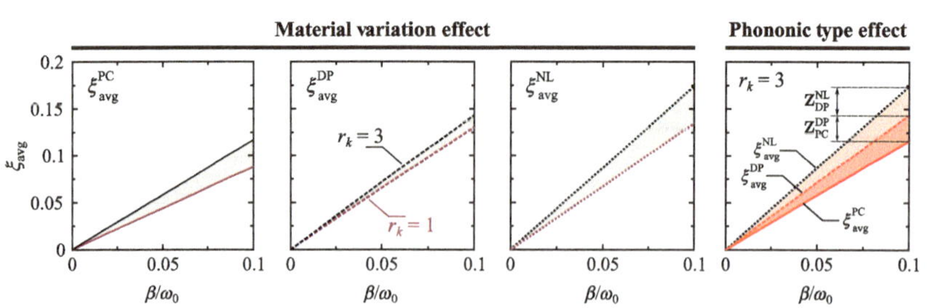

Fig. 16 Effect of material properties (*left*) and phononic material type (*right*) on metadamping for $r_c = 1$.

and r_c are then generated. For a given (constant) amount of prescribed damping η, the two damping coefficients c_1 and c_2 may be computed as

$$c_1 = \frac{\eta}{(1 + r_c)} \tag{41a}$$

$$c_2 = \frac{\eta}{(1 + 1/r_c)} \tag{41b}$$

For $\eta = 15$ and maintaining $\omega_0^{\mathrm{PC}} = 100$ throughout, Fig. 17 displays the relative metadamping differences $\mathbf{Z}_{\mathrm{PC}}^{\mathrm{DP}}$, $\mathbf{Z}_{\mathrm{PC}}^{\mathrm{NL}}$, and $\mathbf{Z}_{\mathrm{DP}}^{\mathrm{NL}}$, as well as the corresponding sonic speed as computed from Eq. (31). Recall that the adjusted ω_0^{DP} and ω_0^{NL} can be obtained from Eqs. (38a) and (38c), respectively. These charts show that the metadamping effectiveness is largely affected by the distribution of damping among the different materials of the phononic material, in addition to the material properties (represented by r_k). For instance, in the case of $\mathbf{Z}_{\mathrm{PC}}^{\mathrm{NL}}$, the metadamping enhancement is seen to increase as we move further away from $r_k = 1$. Furthermore, a large metadamping effect accompanies large values of r_k almost consistently irrespective of r_c. For small r_k, however, the metadamping effect only becomes as significant with larger r_c values. Additionally, the maps also show unfavorable regions (shaded area in Fig. 17) such as in the case of $\mathbf{Z}_{\mathrm{DP}}^{\mathrm{NL}}$, where the metadamping in the conventional DP metamaterial outperforms its counterpart in the NL system.

In addition to confirming the metadamping emergence in resonant metamaterials (e.g., the DP and NL systems) when compared to a common benchmark in the form of a PC system, Fig. 17 also gives a physical insight into how the metadamping in the NL system compares to that of a conventional DP metamaterial. The ability of DP metamaterials to exhibit metadamping gradually vanishes as the stiffness of the local resonator becomes very high to the extent that constrains the internal vibrations and consequently renders the internal resonance dynamics ineffective. This can be noticeably observed in the leftmost plot of Fig. 17 where $\mathbf{Z}_{\mathrm{PC}}^{\mathrm{DP}}$ goes to zero at high values of the stiffness

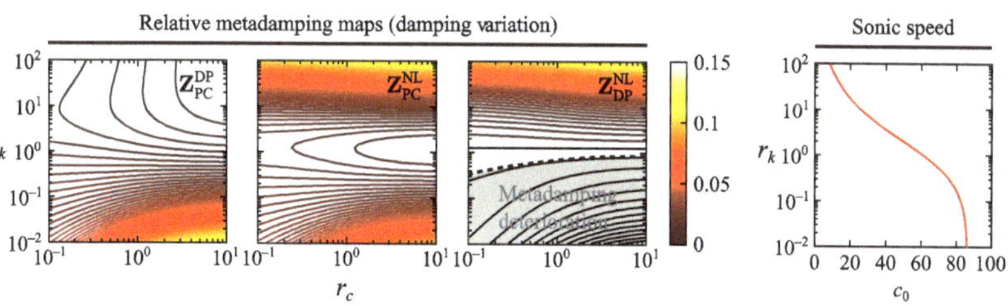

Fig. 17 Contours of the relative metadamping differences $\mathbf{Z}_{\mathrm{PC}}^{\mathrm{DP}}$, $\mathbf{Z}_{\mathrm{PC}}^{\mathrm{NL}}$, and $\mathbf{Z}_{\mathrm{DP}}^{\mathrm{NL}}$ with variations in r_k and r_c. The corresponding sonic speed is provided for reference.

ratio r_k, signaling that the DP and PC systems basically become identical from the standpoint of metadamping. NL metamaterials alleviate this situation by providing a unit cell configuration where the resonator is not completely isolated from the periodic chain and thus mitigates the effect of a high resonator stiffness, all while still maintaining some attributes of an elastic metamaterial as evident by the dispersion diagrams in Fig. 10. As a result, the metadamping in the NL system outperforms that in the DP system and $\mathbf{Z}_{\mathrm{PC}}^{\mathrm{NL}}$ exhibits values greater than zero at higher values of r_k. Once the resonator stiffness is reduced again, the DP metamaterial recovers its metadamping advantage over the NL design, as can be seen in the behavior of $\mathbf{Z}_{\mathrm{DP}}^{\mathrm{NL}}$. The latter corresponds to the region marked "Deterioration" in the rightmost plot of Figure 17.

Analogous to the variations in r_c, the analysis may be extended to depict changes in the metadamping performance with mass and stiffness ratios that span a larger design space. This is shown in Fig. 18 for constant damping coefficients. The damping ratio ξ corresponding to the optical branch at $\kappa a = \pi$ is provided in Fig. 19 for reference. Fig. 19 is used as a precautionary measure to ensure that none of the results displayed in Fig. 18 pertain to an overdamped system (i.e., $\xi > 1$) with a nonoscillatory response which is not of interest in the current analysis. It is also worth noting that some parameter

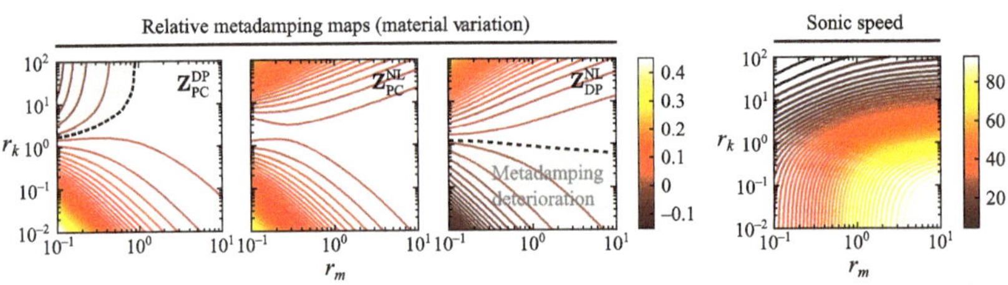

Fig. 18 Contours of the relative metadamping differences $\mathbf{Z}_{\mathrm{PC}}^{\mathrm{DP}}$, $\mathbf{Z}_{\mathrm{PC}}^{\mathrm{NL}}$, and $\mathbf{Z}_{\mathrm{DP}}^{\mathrm{NL}}$ for identical damping coefficients ($c_1 = c_2 = 2$) and variations in r_k and r_m. ω_0^{PC} is kept at 100 for all the considered cases. The corresponding sonic speed is provided for reference.

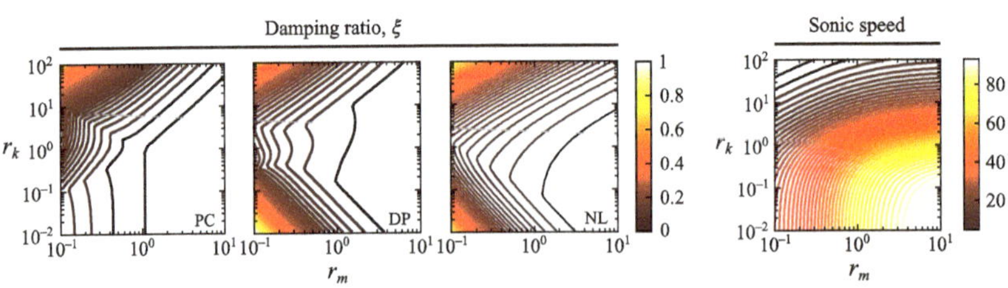

Fig. 19 Damping ratio ξ of the optical dispersion branch at $\kappa a = \pi$ corresponding to the relative metadamping maps provided in Fig. 18.

choices result in negative values for both $\mathbf{Z}_{\mathrm{PC}}^{\mathrm{DP}}$ and $\mathbf{Z}_{\mathrm{DP}}^{\mathrm{NL}}$; an indication of a performance deterioration from a metadamping standpoint. On the contrary, $\mathbf{Z}_{\mathrm{PC}}^{\mathrm{NL}}$ show improvement for the entire design space considered. These charts constitute a comprehensive framework and serve as design guidelines for the type of phononic material, the material properties, as well as the distribution of damping within the same phononic material to yield optimal metadamping levels.

7. EFFECT OF CORE DAMPING MODEL

In the previous sections, the concept of metadamping was presented in the context of viscous damping, i.e., simple dashpot elements. However, accurate and realistic prediction of material damping typically requires different, and often more complex, models. A natural extension of this work is to investigate to what degree the intriguing metadamping attributes established earlier apply to slightly more involved damping models. Means of accounting for damping in structural mechanics include the use of complex elastic moduli (Castanier & Pierre, 1993; Sprik & Wegdam, 1998; Zhang, Liu, Mei, & Liu, 2003) or inertial terms (Langley, 1994b), as well as stand-alone parameters which are functions of velocity (Hussein, 2009b; Hussein & Frazier, 2010). The choice of the damping model usually depends on the application and is critical to minimize discrepancies at the experimental stages (Adhikari & Phani, 2009; Phani & Woodhouse, 2007, 2009). Here, we extend the previous discussion to the viscoelastic regime and specifically the Maxwell element (see Fig. 20) to model the viscoelasticity (Frazier & Hussein, 2015). The mathematical formulation developed in Section 1.2 is slightly modified to accommodate for the differences in the damping mechanism as will be illustrated in the forthcoming section.

7.1 Viscoelasticity: Background and Modeling

Viscoelasticity is a property of a material that simultaneously exhibits viscous and elastic behaviors. As a result, a coupling of the damping and the stiffness properties within the material provides the damping element with

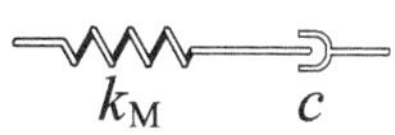

Fig. 20 A schematic of the Maxwell element where the stiffness (conservative) and damping (nonconservative) elements are connected in series.

an energy storage capability. Unlike viscous damping, which is primarily observed in viscous fluids (Woodhouse, 1998), viscoelastic damping is a hereditary model which takes into consideration the past and present history of the dynamical motion whereby dissipation is attributed to a phase shift between the displacement and stress fields (Bert, 1973; Nouh, Aldraihem, & Baz, 2015). Owing to the nature of the viscoelasticity, the mathematical treatment of the viscoelastic model includes internal states $\hat{\mathbf{p}}_j$ which are computed using a convolution integral over an exponentially decaying kernel function $G(t) = \mu_{1,j} e^{-\mu_{2,j} t}$ (where $\mu_{1,j}$ and $\mu_{2,j}$ are the relaxation pairs) capturing the entire motion up to and including the current state (Wagner & Adhikari, 2003). For an equal pair of relaxation parameters, i.e., $\mu_j = \mu_{1,j} = \mu_{2,j}$, the j-th internal state is given by the following convolution integral (Frazier & Hussein, 2015)

$$\hat{\mathbf{p}}_j = \int_0^t \mu_j e^{-\mu_j (t-\tau)} \dot{\hat{\mathbf{u}}}(\tau) \mathrm{d}\tau \tag{42}$$

and, hence, Eq. (5) is rewritten as

$$\hat{\mathbf{M}} \ddot{\hat{\mathbf{u}}} + \sum_{j=1}^{\ell} \hat{\mathbf{C}}_j \hat{\mathbf{p}}_j + \hat{\mathbf{K}} \hat{\mathbf{u}} = \mathbf{0} \tag{43}$$

where ℓ is the total number of internal states. To attain the time derivative of $\hat{\mathbf{p}}_j$, the Leibniz integral rule is applied to Eq. (42) resulting in

$$\dot{\hat{\mathbf{p}}}_j = -\int_0^t \mu_j^2 e^{-\mu_j (t-\tau)} \dot{\hat{\mathbf{u}}}(\tau) \mathrm{d}\tau + \mu_j \dot{\hat{\mathbf{u}}} = \mu_j \left[\dot{\hat{\mathbf{u}}} - \hat{\mathbf{p}}_j \right] \tag{44}$$

following which, substituting back into Eq. (43) gives

$$\hat{\mathbf{M}} \ddot{\hat{\mathbf{u}}} + \sum_{j=1}^{\ell} \hat{\mathbf{C}}_j \left[\dot{\hat{\mathbf{u}}} - \frac{1}{\mu_j} \dot{\hat{\mathbf{p}}}_j \right] + \hat{\mathbf{K}} \hat{\mathbf{u}} = \mathbf{0} \tag{45}$$

Eq. (45) in its current form cannot be readily used in the state-space representation presented earlier since it would result in nonsquare matrices. Premultiplying Eq. (44) by $\frac{1}{\mu_j^2} \hat{\mathbf{C}}_j$ and rearranging yields

$$\frac{1}{\mu_j} \hat{\mathbf{C}}_j \hat{\mathbf{p}}_j - \frac{1}{\mu_j^2} \hat{\mathbf{C}}_j \dot{\hat{\mathbf{p}}}_j - \frac{1}{\mu_j} \hat{\mathbf{C}}_j \dot{\hat{\mathbf{u}}} = \mathbf{0} \tag{46}$$

Plugging Eq. (46) into Eq. (45), the state-space matrices $\hat{\mathbf{A}}$ and $\hat{\mathbf{B}}$ can now be cast as

$$\hat{\mathbf{A}} = \begin{bmatrix} \mathbf{0} & \hat{\mathbf{M}} & \mathbf{0} & \mathbf{0} & \cdots & \mathbf{0} \\ \hat{\mathbf{M}} & \sum_{j=1}^{\ell} \hat{\mathbf{C}}_j & -\frac{1}{\mu_1}\hat{\mathbf{C}}_1 & -\frac{1}{\mu_2}\hat{\mathbf{C}}_2 & \cdots & -\frac{1}{\mu_\ell}\hat{\mathbf{C}}_\ell \\ \mathbf{0} & -\frac{1}{\mu_1}\hat{\mathbf{C}}_1 & \frac{1}{\mu_1^2}\hat{\mathbf{C}}_1 & \mathbf{0} & \cdots & \mathbf{0} \\ \mathbf{0} & -\frac{1}{\mu_2}\hat{\mathbf{C}}_2 & \mathbf{0} & \frac{1}{\mu_2^2}\hat{\mathbf{C}}_2 & \cdots & \mathbf{0} \\ \vdots & \vdots & \vdots & \vdots & \ddots & \vdots \\ \mathbf{0} & -\frac{1}{\mu_\ell}\hat{\mathbf{C}}_\ell & \mathbf{0} & \mathbf{0} & \cdots & \frac{1}{\mu_\ell^2}\hat{\mathbf{C}}_\ell \end{bmatrix} \tag{47a}$$

$$\hat{\mathbf{B}} = \mathbf{diag}\left[-\hat{\mathbf{M}} \;\; \hat{\mathbf{K}} \;\; \frac{1}{\mu_1}\hat{\mathbf{C}}_1 \;\; \frac{1}{\mu_2}\hat{\mathbf{C}}_2 \;\; \cdots \;\; \frac{1}{\mu_\ell}\hat{\mathbf{C}}_\ell\right] \tag{47b}$$

where $\hat{\mathbf{y}}^{\mathrm{T}} = \left[\dot{\hat{\mathbf{u}}} \;\; \hat{\mathbf{u}} \;\; \hat{\mathbf{p}}_1 \;\; \hat{\mathbf{p}}_2 \;\; \ldots \;\; \hat{\mathbf{p}}_\ell\right]$ is the new state vector. While multiple internal states $\hat{\mathbf{p}}_j$ more accurately describe the viscoelastic working mechanism, the focus of the present analysis is to highlight the differences between the viscoelastic and viscous damping and, therefore, a single kernel function (i.e., $\ell = 1$) is sufficient to describe the viscoelastic phenomenon; Eq. (8) now reads

$$\left(\lambda \begin{bmatrix} \mathbf{0} & \hat{\mathbf{M}} & \mathbf{0} \\ \hat{\mathbf{M}} & \hat{\mathbf{C}} & -\frac{1}{\mu}\hat{\mathbf{C}} \\ \mathbf{0} & -\frac{1}{\mu}\hat{\mathbf{C}} & \frac{1}{\mu^2}\hat{\mathbf{C}} \end{bmatrix} + \begin{bmatrix} -\hat{\mathbf{M}} & \mathbf{0} & \mathbf{0} \\ \mathbf{0} & \hat{\mathbf{K}} & \mathbf{0} \\ \mathbf{0} & \mathbf{0} & \frac{1}{\mu}\hat{\mathbf{C}} \end{bmatrix}\right)\hat{\mathbf{y}} = \mathbf{0}. \tag{48}$$

Upon computing the eigenvalues of Eq. (48), the real nonoscillatory modes corresponding to the internal states can be found. The remaining modes are complex pairs which dictate wave propagation in the dissipative structure, as long as that the prescribed damping intensity β does not render the system overdamped (i.e., $\xi > 1$). It is worth noting that in the extreme cases where $\mu \rightarrow 0$ and $\mu \rightarrow \infty$, Eq. (48) recovers the undamped and viscously damped state matrices, respectively. With $\mu \rightarrow 0$, and since $k_{\mathrm{M}} \propto \mu$, the spring in Fig. 20 softens and eventually disengages. On the other hand, for $\mu \rightarrow \infty$, the spring stiffens and effectively becomes a rigid connection. As a consequence, most of the energy is dissipated in the damping element c.

7.2 Metadamping in Viscoelastic Phononic Materials

7.2.1 Viscous-to-Viscoelastic Transition

In this section, we revisit the example in Fig. 11 and apply the viscoelastic model for the same parameters to capture the transition between viscous to viscoelastic damping. If the relaxation parameter $\mu = 300$ is used to generate the viscoelastic counterpart of Fig. 11, the result will be as depicted in Fig. 21 where the undamped and viscously damped cases are also plotted for reference. In all the shown cases, the optical branches as expected demonstrate higher tendency to deviate from the undamped case given the higher frequency range. The acoustical branches remain nearly unchanged. The introduction of viscoelastic damping tends to increase the damped frequency since some of the energy in the Maxwell element is conserved (unlike the viscous model which effectively decreases the damped frequency). This increase in the damped frequency of the optical branch widens the band gap, an effect that is more pronounced in the case of the PC. Since part of the energy is stored in the viscoelastic model, we expect the damping ratios of the viscoelastic systems to be lower than those of the viscous ones across all the phononic materials considered. This can be noticeably observed in the lower panel of Fig. 21.

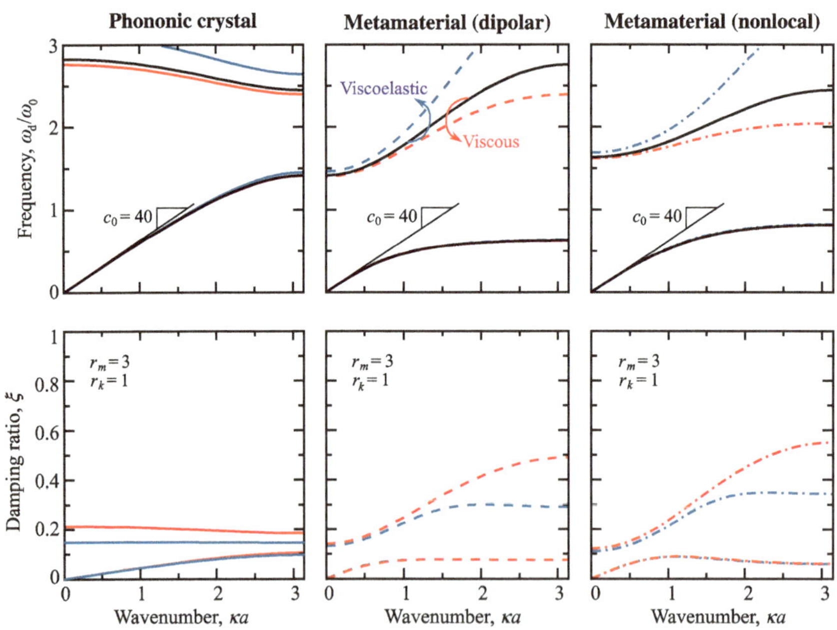

Fig. 21 Comparison of the dispersion curves and corresponding damping ratios for the PC, DP, and NL systems under viscous and viscoelastic models.

7.2.2 Band-Gap Limits

The evolution of the band-gap limits as well as the band-gap width $\Delta\omega_d$ for all the different viscoelastic phononic materials is shown in Fig. 22 by the solid curves. The dashed lines denote the viscous and undamped models and are shown for comparison. The narrowest band gaps take place when the phononic material has fully developed into the viscous damping regime. During the transition, the band-gap width for the PC peaks in the vicinity of $\mu = 200$, while for the NL and DP metamaterials, the largest band-gap width occurs around $\mu = 120$. These peaks happen as the energy storage capability in the Maxwell element reaches its maximum value thus overshadowing the amount of dissipated energy (Frazier & Hussein, 2015).

Despite the increase in the band-gap width, we anticipate the metadamping emergence in the viscoelastic phononic materials to deteriorate. This can also be inferred from the damping ratio ξ comparisons displayed in Fig. 21. Following a similar procedure to that used to generate Fig. 13, and using $\mu = 300$, the

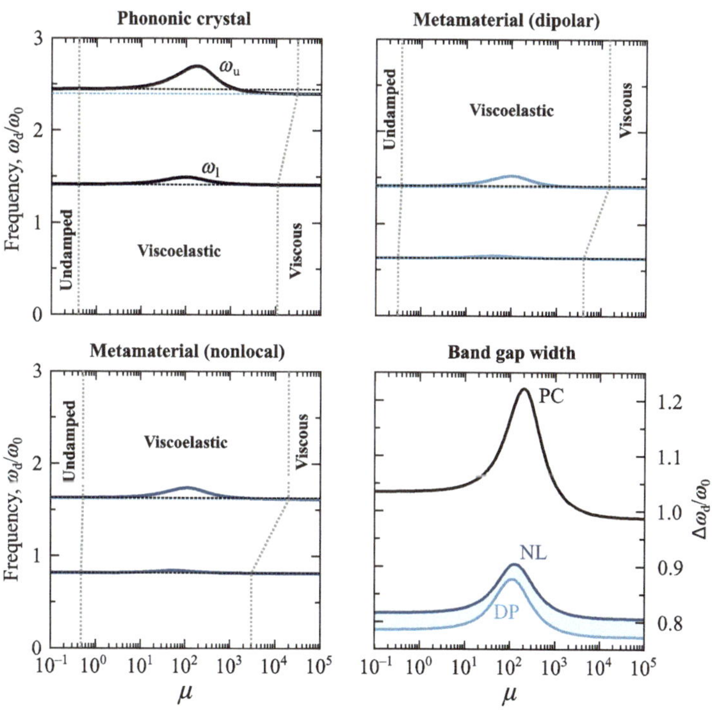

Fig. 22 Evolution of band-gap limits for the PC, DP, and NL systems as a function of the relaxation parameter μ. The figure illustrates the transition lines between the undamped, viscoelastic and viscous regimes. *Black* and *blue dashed lines* represent the band-gap limits of the undamped and the viscous models, respectively.

sonic speed c_0 is swept across a range of values and the corresponding metadamping ξ_{avg} for the viscoelastic phononic materials is evaluated. The relative metadamping performances of the PC, DP, and NL systems are shown in Fig. 23A. Further, a comparison of the emergent metadamping between the viscoelastic and the viscous systems is presented in Fig. 23B–D. Examining these patterns, it can be seen that the sensitivity of the metadamping to the damping model decreases at lower sonic speeds where, for the given relaxation parameter, the viscoelastic model becomes nearly viscous.

Finally, the evolution of metadamping with the change in the damping regimes (i.e., the degree of the hereditary model) is computed as a function of μ for the different phononic materials in Fig. 24. The metadamping increases as μ increases for all the phononic materials up until a certain degree of saturation as the systems completely transition into the viscous regime. In contrast to the PC, the DP and NL metamaterials exhibit a ξ_{avg} peak in the neighborhood of the transition line between the viscous and viscoelastic models. Such unique traits which manifest themselves from the integration

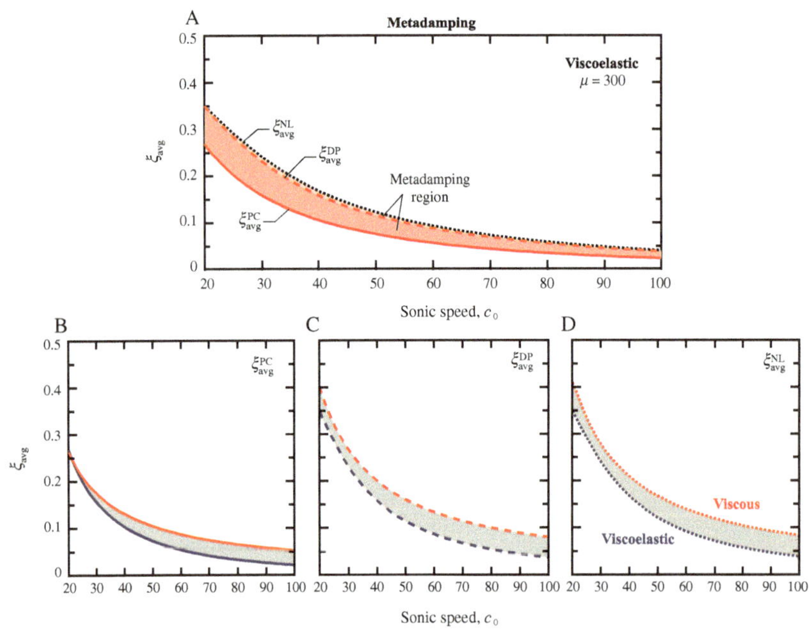

Fig. 23 (A) Metadamping emergence in NL and DP metamaterials with viscoelastic damping ($\mu = 300$). (B)–(D) Quantitative comparisons between the viscoelastic and the viscous metadamping curves for the PC, DP, and NL systems, respectively.

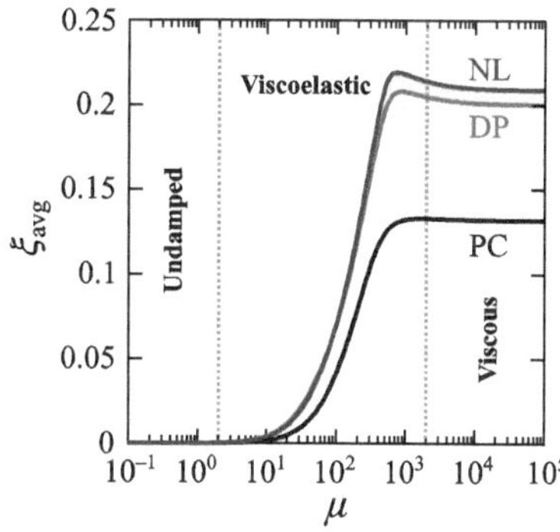

Fig. 24 Metadamping in the PC, DP, and NL systems as a function of the relaxation parameters μ. Transition between the different regimes is illustrated by the *dotted lines*.

of the material damping with the different phononic materials open up new avenues in material designs for optimal wave absorption as well as broadband attenuation performances.

8. EXPERIMENTAL VALIDATION OF METADAMPING

In the previous sections, the concept of metadamping has been presented and analyzed in the context of lumped mass-spring-dashpot systems. Although these simple models provide a clear demonstration of the concept, with the added advantage of enabling closed-form analytical solutions, a more elaborate model is needed to examine suitability for real-world applications and to enable proper experimental validation.

In this section, we examine metadamping in an experimental configuration, and a corresponding numerical model, in the form of an extended all-aluminum beam with pillars periodically standing out—along the axial axis—from one of the beam's surfaces (see Figs. 25 and 26) (Bacquet & Hussein, 2018). The pillars serve the role of the resonating substructures; they are shaped by milling to ensure seamless connectivity. Our choice of a medium with one-dimensional periodicity is only for ease of exposition as the underlying dynamical behavior we are interested in takes place for any dimension and is in fact independent of periodicity (Achaoui, Laude, Benchabane, & Khelif, 2013). This investigation consists of two parts. In the first part, we use experiments and corresponding finite-element simulations applied to a four-celled finite-sized version of the pillared

Fig. 25 Photograph of the two experimental structures: the unpillared beam (*left*) and the pillared beam (*right*).

Fig. 26 Numerical models of the two experimental structures: the unpillared beam (*left*) and the pillared beam (*right*).

beam to provide evidence of metadamping. In the second part, we provide a correlation of the observed behavior with the dispersion and damping ratio diagrams for the unit cell from which the pillared beam structure is formed.

8.1 Pillared Beam: One-Dimensional Metamaterial

We consider the following two test structures composed entirely from aluminum: a regular beam (unpillared) and a metamaterial beam (pillared). The unpillared beam's dimensions are 32 × 1 × 1 inches. The pillared beam has the same underlying beam base, but is augmented with four squares pillars (0.5 × 0.5 × 2 inches) which are periodically arranged along the axial direction. These pillars act as the local resonators; there is one pillar per unit cell. Fig. 25 shows a photograph of both structures in the laboratory, and Fig. 26 depicts the corresponding numerical models. Throughout this section, the color blue will represent the unpillared beam and the color red will represent the pillared beam.

8.1.1 Experimental Setup

Each of the beam structures is suspended using nylon cords in order to simulate free-free boundary conditions. An accelerometer (PCB WJ35C65) is attached at the center point of the cross section on one of the ends of the beam. Impulsive excitations are applied with an impact hammer (PCB 086C02) at a point located at the center of the opposite cross section, such that only the longitudinal modes are excited and measured. The measurements are collected with a NI-DAQ 9234 data acquisition system (the frequency rate is set to 25.6 kHz and the sample time is 5 s). Five time series are recorded and averaged for each beam. The inertance spectrum is obtained by postprocessing the time data using a commercial software package (MATLAB®, The MathWorks Inc., Natick, MA, United States). Fig. 27A shows the experimental frequency response functions (FRF) obtained by this process. The reader is referred to the review article Hussein et al. (2014) and the recent article Al Ba'ba'a, Attarzadeh, and Nouh (2017) for examples of other experimental setups for phononic systems.

8.1.2 Numerical Model

The two beam structures are numerically modeled with the following prescribed material properties: density of $\rho = 2700$ kg/m^3, Young's modulus of $E = 68.9$ GPa, and Poisson's ratio of $\nu = 0.33$. Using the finite-element (FE) method for spatial discretization, we adopt a viscoelastic damping model of the form (Frazier & Hussein, 2015; Hussein & Frazier, 2013a; Wagner & Adhikari, 2003)

$$\mathbf{M}\ddot{\mathbf{u}}(t) + \int_{k=1}^{t} \mu e^{-\mu(t-\tau)} \mathbf{C}\dot{\mathbf{u}}(\tau) d\tau + \mathbf{K}\mathbf{u}(t) = \mathbf{f}(t), \tag{49}$$

where, for simplicity, the FE damping matrix $\mathbf{C}$ is assumed to be proportional to the FE mass $\mathbf{M}$ and stiffness $\mathbf{K}$ matrices, such that $\mathbf{C} = p\mathbf{M} + q\mathbf{K}$. The FE model consists of three-dimensional 8-node brick FE (total number of elements in the unit cell are 4096 for the unpillared beam and 4352 for the pillared beam). The parameters p and q, as well as a relaxation parameter μ, are determined by a unique experimental material–structure curve-fitting procedure described in Bacquet and Hussein (2018), which has provided us with the values $p = 22$, $q = 2.2 \times 10^{-7}$, and $\mu = 10^4$. Information on the FE implementation is available in Hussein (2009 a); and information on the viscoelastic

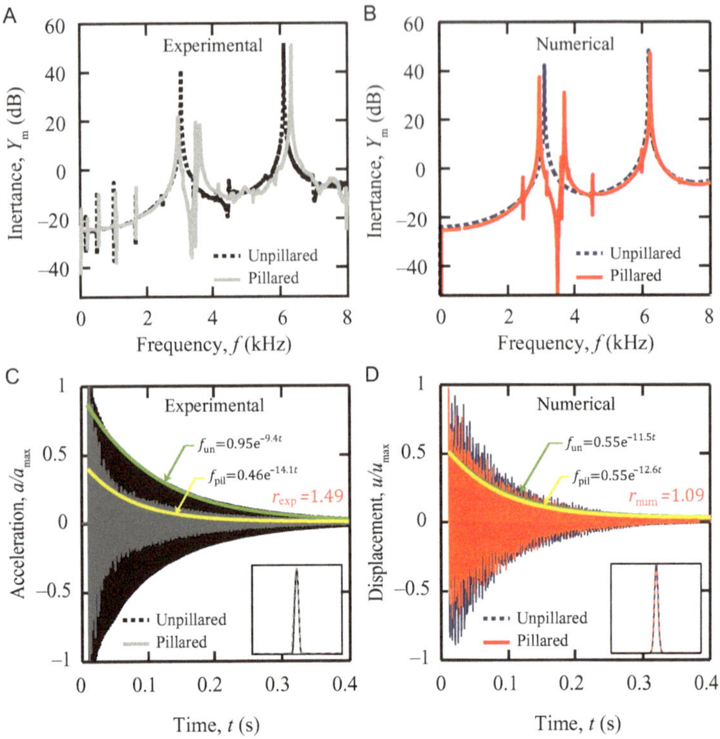

Fig. 27 (A) Experimental and (B) numerical frequency response functions for the unpillared and for the pillared beams. (C) Experimental and (D) and numerical temporal responses for the two beam structures. The *green* and *yellow curves* are the curve-fitted exponential functions for the unpillared and pillared time signals, respectively, to determine their time decay rates. The form of impulse excitations are shown in the insets; these are obtained by a hammer impact in the experimental case, and mathematically synthesized in the numerical case.

model adopted, and the application of Bloch's theorem to that model, are found in Frazier and Hussein (2015) and Hussein and Frazier (2013a). The numerical FRFs (Figs. 27B) are shown to agree very well with their experimental counterparts (Fig. 27A).

8.2 Evidence of Metadamping

To determine the presence of metadamping, we seek to show that the pillared beam exhibits stronger temporal attenuation than the unpillared beam. This may be demonstrated for a finite beam by examining the temporal response due to impulse excitation, and for an infinite beam by examining the damping ratio diagram obtained from a Bloch analysis on the unit cell.

8.2.1 Finite Structure

The experimental temporal response for the same test structures investigated in Fig. 27A to a particular profile of impulse excitation is shown in Fig. 27C. To identify the presence of metadamping and quantify its intensity, we compare the rate of time decay of the pillared beam and compare it to that of the unpillared beam. The procedure is as follows. First, the first derivative of the displacement response is evaluated. Then, the maxima, which are the points where the derivative changes sign, are extracted. Exponential functions of the form $f(t) = ae^{-bt}$ are then curve-fitted to these extracted response peaks, where the exponential decay constant b provides a direct measure of the degree of dissipation.

With this approach, a metric for metadamping is defined as the ratio $r = b_{\text{pil}}/b_{\text{un}}$, where "pil" and "un" denote pillared and unpillared, respectively. A ratio greater than unity signifies positive metadamping, i.e., the time response of the pillared beam decays faster than that of the unpillared beam. Conversely, a ratio less than unity indicates negative metadamping [not covered here; see Frazier and Hussein (2015) and Bacquet and Hussein (2018)]. From the results of Fig. 27C, we report an experimental metadamping ratio $r_{\text{exp}} = 1.49$ which is indicative of nearly 50% positive metadamping, i.e., the pillared beam exhibits 50% higher dissipation than the unpillared beam for this particular form of excitation.

A similar analysis is conducted using our FE beam models by implementing a time-integration scheme particularly suited for exponentially damped systems (Adhikari & Wagner, 2004). In order to accurately replicate the experimental setup, the initial displacement is modeled as a Gaussian excitation whose parameters a and b are selected such that it matches the experimental impulse:

$$u_0(t) = e^{-\frac{(t-a)^2}{2b^2}}, \tag{50}$$

where $a = 0.01$ and $b = 8 \times 10^{-5}$. The time response of the beams are computed for $t = 0.4$ s with a time step of $\Delta t = 3 \times 10^{-5}$ s. The results, which are shown in Fig. 27D, give us a numerical metadamping ratio of $r_{\text{num}} = 1.09$. While the degree of the predicted metadamping is lower than the experimental results, this confirms that the pillared beam exhibits higher dissipation for the applied excitation profile.

8.2.2 Infinite Material

To elucidate the metadamping phenomenon, and what causes it in the finite structure as we observed by brute-force simulations, we revisit the problem and examine it from the material point of view. This is done by performing a dispersion analysis on the unit cells (both the unpillared and pillared) and comparing their damping ratio diagrams. The free equation of motion for each unit cell is given by Eq. (49) with the force term on the right-hand side set to zero. The dispersion band structure and corresponding damping ratio diagram are obtained by assuming a Bloch solution of the form $\boldsymbol{u}(t) = \widetilde{\boldsymbol{U}}(\kappa)e^{i\kappa x + \lambda t}$ and substituting it into Eq. (49). Upon transformation to state space, we obtain the following matrix eigenvalue problem [details of the derivation are available in Hussein and Frazier (2010, 2013a) and Frazier and Hussein (2015)]:

$$\left(\lambda\begin{bmatrix} \mathbf{0} & \mathbf{M} & \mathbf{0} \\ \mathbf{M} & \mathbf{C}(\kappa) & -\dfrac{1}{\mu}\mathbf{C}(\kappa) \\ \mathbf{0} & -\dfrac{1}{\mu}\mathbf{C}(\kappa) & \dfrac{1}{\mu^2}\mathbf{C}(\kappa) \end{bmatrix} + \begin{bmatrix} -\mathbf{M} & \mathbf{0} & \mathbf{0} \\ \mathbf{0} & \mathbf{K}(\kappa) & \mathbf{0} \\ \mathbf{0} & \mathbf{0} & \dfrac{1}{\mu}\mathbf{C}(\kappa) \end{bmatrix}\right)\begin{bmatrix} \dot{\boldsymbol{u}} \\ \boldsymbol{u} \\ \boldsymbol{p} \end{bmatrix} = \mathbf{0}, \tag{51}$$

This problem is solved for prescribed and real wavenumbers κ spanning the first Brillouin zone ($0 \leq \kappa a \leq \pi$) which gives us wavenumber-dependent, complex eigenvalues $\lambda_s(\kappa)$ in the form

$$\lambda_s(\kappa) = -\xi_s(\kappa)\omega_s(\kappa) \pm \mathrm{i}\omega_{d_s}(\kappa), \quad s = 1, \ldots, n, \tag{52}$$

where n is the total number of modes. The imaginary part ω_{d_s} represents the damped frequency corresponding to the s-th Bloch mode, and the real part is the product of the wavenumber-dependent damping ratio ξ_s and, in the case of Rayleigh (proportional) damping, the undamped frequency ω_s. The dispersion diagram for both beam structures is shown in Fig. 28A and the corresponding damping ratio diagram is plotted in Fig. 28B (we only show the relevant portions of each diagram). The latter diagram provides us with the level of dissipation that each Bloch mode exhibits.

We observe in the dispersion diagram (Fig. 28A) two curves cutting through horizontally; these represent local resonances associated with pillar motion. These curves couple with the underlying dispersion curves describing wave motion along the base beam structure and cause the hybridization phenomenon that is characteristic of locally resonant metamaterials, as we observed in the earlier sections dealing with lumped-parameter models. The pillared beam dispersion curves clearly contrast with those of unpillared beam as the latter curves do not exhibit any local resonances. Since our earlier experimental focus has been in longitudinal displacements (and accelerations), we specifically highlight the longitudinal modes and shade the other

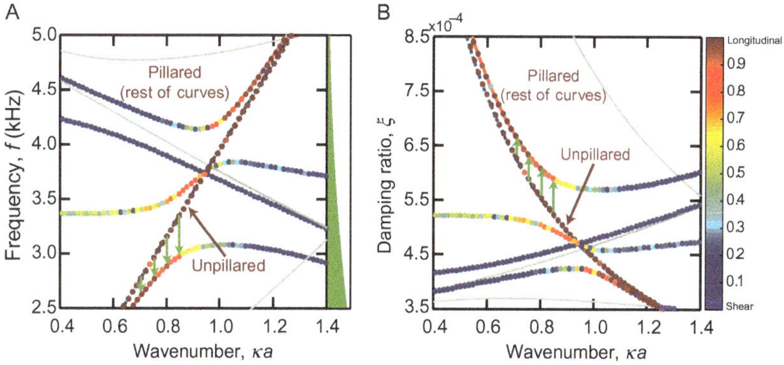

Fig. 28 (A) Dispersion and (B) damping ratio diagrams with a close-up view of the hybridization region. The greens arrows illustrate the metadamping effect due to the addition of the pillars. The *color bar* represents the degree of longitudinal polarization in each mode. The frequency content of the impulse excitation is plotted and shaded in *green* on the far end of (A) demonstrating overlap with metadamping region; most of the excitation frequency content falls within the $2500 \leq f \leq 3250$ Hz range.

modes in gray. To rigorously identify which modes are longitudinal in the dispersion diagram, we first sort the Bloch modes using an algorithm that checks continuation of group velocities [which is a measure of the degree of orthogonality between two vectors (Allemang, 2003)], and then calculate the degree of longitudinal polarization following the method described in Achaoui, Khelif, Benchabane, and Laude (2010). Polarization values range from 0 (pure shear) to 1 (pure longitudinal).

The key feature in Fig. 28 is the green arrows that illustrate the transition of the dispersion curves and damping ratios due to the inclusion of the pillar in the unit cell. In the damping ratio diagram in particular, these arrows indicate an increase in the level of dissipation. And since this takes place in the frequency range of the excitation, the finite structure formed from this constituent cell experiences increased dissipation, or metadamping, as we observed in Figs. 27C and D.

8.3 Effect of Added Mass on Dissipation

Finally, it is important to ensure that the added mass of the pillar is not a factor in the generation of metadamping. The unpillared beam has a volume of $V_{\text{un}} = 32$ in^3 whereas the pillared beam has a volume of $V_{\text{pil}} = 34$ in^3. Therefore, to show that the metadamping is due to the local resonance

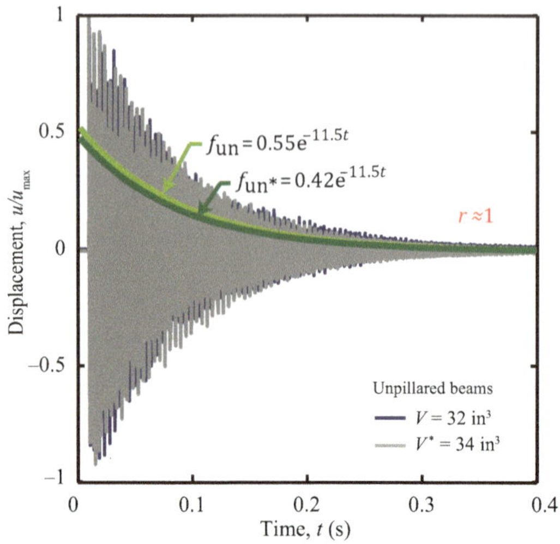

Fig. 29 Numerical time responses for the two unpillared beams, with volumes $V = 32$ in^3 (*blue*) and $V^* = 34$ in^3 (*gray*), respectively. The *light* and *dark green curves* correspond to exponential curves that were fitted to the unpillared and unpillared volume time signals, respectively, to determine their time decays.

phenomena only and not due to the addition of extra mass that exhibits material damping, we perform a similar analysis as that of Fig. 28D on an unpillared beam that has the same volume as the pillared beam. The numerical time responses and their curve-fitted exponential functions are shown in Fig. 29. The metadamping ratio, here redefined as $r = b_{un^*}/b_{un}$, is shown to equal one; this confirms our hypothesis that metadamping emerges from the presence of local resonance and not from the addition of more damped material.

9. CONCLUSIONS

Metadamping is a resonance-enabled intrinsic damping emergence phenomenon whereby the level of dissipation may be enhanced (or reduced) in a metamaterial compared to a statically equivalent material with the same long-wave speed and type and quantity of prescribed damping. While elastic metamaterials are known to provide strong spatial attenuation inside band gaps, metadamping enables the additional trait of strong temporal attenuation across relatively broad frequency ranges. The phenomenon takes place at regions in the wavenumber-dependent damping ratio diagram where the dissipation is shown to increase due to the inclusions of the resonances.

In this chapter, we have provided a technical review of the concept for different types of resonances, namely local dipole, local monopole, and non-local. It is shown that the quest for high metadamping in resonant metamaterials becomes a trade-off between seeking the absolute highest metadamping (provided by DP metamaterials) or a considerable metadamping effect across a wider choice of resonator stiffnesses (provided by NL metamaterials). A parametric study and an examination of the effects of viscous vs nonviscous damping are also provided. The chapter concludes with a real-life experimental investigation of metadamping in a pillared beam, where we show matching results with both simulations and band-structure theory. Future research may take the phenomenon of metadamping to the microscale and to waves driven at prescribed frequency.

ACKNOWLEDGMENTS

MIH acknowledges the National Science Foundation for its partial support of this work through the NSF CAREER Grant No. 1254931 and Grant No. 1538596. MIH and CLB thank Professor Srikanth Phani for fruitful discussions. MN acknowledges the National Science Foundation for its support of this work through Grant No. 1647744.

REFERENCES

Achaoui, Y., Khelif, A., Benchabane, S., & Laude, V. (2010). Polarization state and level repulsion in two-dimensional phononic crystals and waveguides in the presence of material anisotropy. *Journal of Physics D: Applied Physics*, *43*(18), 185401. https://doi.org/10.1088/0022-3727/43/18/185401. http://stacks.iop.org/0022-3727/43/i=18/a=185401?key=crossref.4cfb160c209188b55ba1ab0b245fa8bd. (updated 2017-11-17).

Achaoui, Y., Laude, V., Benchabane, S., & Khelif, A. (2013). Local resonances in phononic crystals and in random arrangements of pillars on a surface. *Journal of Applied Physics*, *114*, 104503.

Adhikari, S., & Phani, A. S. (2009). Experimental identification of generalized proportional viscous damping matrix. *Journal of Vibration and Acoustics-Transactions of the ASME*, *131*(1), 011008.

Adhikari, S., & Wagner, N. (2004). Direct time-domain integration method for exponentially damped linear systems. *Computers & Structures*, *82*(29–30), 2453–2461. https://doi.org/10.1016/j.compstruc.2004.08.004. http://www.sciencedirect.com/science/article/pii/S0045794904003074. (updated 2016-06-02).

Al Ba'ba'a, H., DePauw, D., Singh, T., & Nouh, M. (2018). Dispersion transitions and pole-zero characteristics of finite inertially amplified acoustic metamaterials. *Journal of Applied Physics*, *123*, 105106.

Al Ba'ba'a, H., & Nouh, M. (2017). An investigation of vibrational power flow in one-dimensional dissipative phononic structures. *Journal of Vibration and Acoustics-Transactions of the ASME*, *139*(2), 021003–021010.

Al Ba'ba'a, H. B., Attarzadeh, M. A., & Nouh, M. (2017). Experimental evaluation of structural intensity in two dimensional plate-type locally resonant elastic metamaterials. *Journal of Applied Mechanics-Transactions of the ASME*, *122*, 341–354.

Al Ba'ba'a, H. B., & Nouh, M. (2017). Mechanics of longitudinal and flexural locally resonant elastic metamaterials using a structural power flow approach. *International Journal of Mechanical Sciences*, *122*, 341–354.

Allemang, R. J. (2003). The modal assurance criterion-twenty years of use and abuse. *Sound and Vibration*, *37*, 14–20.

Antoniadis, I., Chronopoulos, D., Spitas, V., & Koulocheris, D. (2015). Hyper-damping properties of a stiff and stable linear oscillator with a negative stiffness element. *Journal of Sound and Vibration*, *346*, 37–52.

Bacquet, C. L., & Hussein, M. I. (2018). *Dissipation engineering in metamaterials by localized structural dynamics*. arXiv:1809.04509 [physics.app-ph].

Beli, D., Arruda, J. R. F., & Ruzzene, M. (2018). Wave propagation in elastic metamaterial beams and plates with interconnected resonators. *International Journal of Solids and Structures*, *139–140*, 105–120.

Bert, C. W. (1973). Material damping: An introductory review of mathematic measures and experimental technique. *Journal of Sound and Vibration*, *29*(2), 129–153.

Bilal, O. R., & Hussein, M. I. (2013). Trampoline metamaterial: Local resonance enhancement by springboards. *Applied Physics Letters*, *103*(11), 111901.

Bloch, F. (1929). Über die quantenmechanik der elektronen in kristallgittern. *Zeitschrift für Physik*, *52*(7–8), 555–600.

Brun, M., Guenneau, S., & Movchan, A. B. (2009). Achieving control of in-plane elastic waves. *Applied Physics Letters*, *94*(6), 061903.

Cassedy, E. S. (1967). Dispersion relations in time-space periodic media part II—Unstable interactions. *Proceedings of the IEEE*, *55*(7), 1154–1168.

Castanier, M. P., & Pierre, C. (1993). Individual and interactive mechanisms for localization and dissipation in a mono-coupled nearly-periodic structure. *Journal of Sound and Vibration*, *168*(3), 479–505.

Chen, Y., Hu, G., & Huang, G. (2017). A hybrid elastic metamaterial with negative mass density and tunable bending stiffness. *Journal of the Mechanics and Physics of Solids*, *105*, 179–198.

Christensen, J., Kadic, M., Kraft, O., & Wegener, M. (2015). Vibrant times for mechanical metamaterials. *MRS Communications*, *5*(3), 453–462. https://doi.org/10.1557/mrc.2015.51.

Collet, M., Ouisse, M., Ruzzene, M., & Ichchou, M. N. (2011). Floquet-Bloch decomposition for the computation of dispersion of two-dimensional periodic, damped mechanical systems. *International Journal of Solids and Structures*, *48*(20), 2837–2848.

Crandall, S. H. (1970). Role of damping in vibration theory. *Journal of Sound and Vibration*, *11*, 3–18.

Cummer, S. A., & Schurig, D. (2007). One path to acoustic cloaking. *New Journal of Physics*, *9*(3), 45.

DePauw, D., Al Ba'ba'a, H., & Nouh, M. (2018). Metadamping and energy dissipation enhancement via hybrid phononic resonators. *Extreme Mechanics Letters*, *18*, 36–44.

Deymier, P. A. (2013). *Acoustic metamaterials and phononic crystals*. Heidelberg: Springer, (Ed.).

Farzbod, F., & Leamy, M. J. (2011). Analysis of Bloch's method in structures with energy dissipation. *Journal of Vibration and Acoustics-Transactions of the ASME*, *133*(5), 051010.

Floquet, G. (1883). Sur les équations différentielles linéaries à coefficients périodiques. *Annales de l'Ecole Normale Supérieur*, *12*, 47–88.

Frazier, M. J., & Hussein, M. I. (2015). Viscous-to-viscoelastic transition in phononic crystal and metamaterial band structures. *The Journal of the Acoustical Society of America*, *138*(5), 3169–3180.

Frazier, M. J., & Hussein, M. I. (2016). Generalized Bloch's theorem for viscous metamaterials: Dispersion and effective properties based on frequencies and wavenumbers that are simultaneously complex. *Comptes Rendus Physique*, *17*(5), 565–577. https://doi.org/10.1016/j.crhy.2016.02.009.

Hu, G., Tang, L., Das, R., Gao, S., & Liu, H. (2017). Acoustic metamaterials with coupled local resonators for broadband vibration suppression. *AIP Advances*, *7*(2), 025211.

Hussein, M. I. (2009a). Reduced Bloch mode expansion for periodic media band structure calculations. *Proceedings of the Royal Society A*, *465*(2109), 2825–2848.
Hussein, M. I. (2009). Theory of damped Bloch waves in elastic media. *Physical Review B*, *80*(21), 212301.
Hussein, M. I., Biringen, S., Bilal, O. R., & Kucala, A. (2015). Flow stabilization by subsurface phonons. *Proceedings of the Royal Society A: Mathematical, Physical and Engineering Sciences*, *471*(2177), 20140928.
Hussein, M. I., & Frazier, M. J. (2010). Band structure of phononic crystals with general damping. *Journal of Applied Physics*, *108*(9), 093506.
Hussein, M. I., & Frazier, M. J. (2013a). Damped phononic crystals and acoustic metamaterials. In P. A. Deymier (Ed.), *Acoustic metamaterials and phononic crystals* (pp. 201–215). Berlin, Heidelberg: Springer. https://doi.org/10.1007/978-3-642-31232-8_6. Chapter 6 (Ed.).
Hussein, M. I., & Frazier, M. J. (2013b). Metadamping: An emergent phenomenon in dissipative metamaterials. *Journal of Sound and Vibration*, *332*, 4767–4774.
Hussein, M. I., Frazier, M. J., & Abedinnassab, M. H. (2013). Microdynamics of phononic materials. In S. Li & X.-L. Gao (Eds.), *Handbook of micromechanics and nanomechanics*. Pan Stanford Publishing. Chapter 1 (Eds.).
Hussein, M. I., Leamy, M. J., & Ruzzene, M. (2014). Dynamics of phononic materials and structures: Historical origins, recent progress, and future outlook. *Applied Mechanics Reviews-Transactions of the ASME*, *66*(4), 040802.
Khajehtourian, R., & Hussein, M. I. (2014). Dispersion characteristics of a nonlinear elastic metamaterial. *AIP Advances*, *4*(12), 124308.
Khelif, A., & Adibi, A. (2015). *Phononic crystals: Fundamentals and applications*. New York: Springer. (Eds.).
Kohn, W., Krumhansl, J. A., & Lee, E. H. (1972). Variational methods for dispersion relations and elastic properties of composite materials. *Journal of Applied Mechanics-Transactions of the ASME*, *39*(2), 327–336.
Kushwaha, M. S., Halevi, P., Dobrzynski, L., & Djafari-Rouhani, B. (1993). Acoustic band structure of periodic elastic composites. *Physical Review Letters*, *71*(13), 2022–2025.
Langley, R. S. (1994a). On the forced response of one-dimensional periodic structures: Vibration localization by damping. *Journal of Sound and Vibration*, *178*(3), 411–428.
Langley, R. S. (1994b). On the forced response of one-dimensional periodic structures: Vibration localization by damping. *Journal of Sound and Vibration*, *178*(3), 411–428.
Laude, V. (2015). *Phononic crystals: Artifical crystals for sonic, acoustic and elastic waves*. Berlin: De Gruyter.
Lee, E. H., & Yang, W. H. (1973). On waves in composite materials with periodic structure. *SIAM Journal on Applied Mathematics*, *25*(3), 492–499.
Liu, L., & Hussein, M. I. (2012). Wave motion in periodic flexural beams and characterization of the transition between Bragg scattering and local resonance. *Journal of Applied Mechanics-Transactions of the ASME*, *79*(1), 011003.
Liu, Z., Chan, C. T., & Sheng, P. (2005). Analytic model of phononic crystals with local resonances. *Physical Review B*, *71*(1), 014103.
Liu, Z., Zhang, X., Mao, Y., Zhu, Y. Y., Yang, Z., Chan, C. T., & Sheng, P. (2000). Locally resonant sonic materials. *Science*, *289*(5485), 1734–1736.
Mead, D. J. (1973). A general theory of harmonic wave propagation in linear periodic systems with multiple coupling. *Journal of Sound and Vibration*, *27*(2), 235–260.
Merheb, B., Deymier, P. A., Jain, M., Aloshyna-Lesuffleur, M., Mohanty, S., Berker, A., & Greger, R. W. (2008). Elastic and viscoelastic effects in rubber/air acoustic band gap structures: A theoretical and experimental study. *Journal of Applied Physics*, *104*(6), 064913.
Moiseyenko, R. P., & Laude, V. (2011). Material loss influence on the complex band structure and group velocity in phononic crystals. *Physical Review B*, *83*(6), 064301.

Nouh, M., Aldraihem, O., & Baz, A. (2015). Wave propagation in metamaterial plates with periodic local resonances. *Journal of Sound and Vibration*, *341*, 53–73. https://doi.org/10.1016/j.jsv.2014.12.030.

Pendry, J. B. (2000). Negative refraction makes a perfect lens. *Physical Review Letters*, *85*(18), 3966–3969.

Pendry, J. B., Holden, A. J., Robbins, D. J., & Stewart, W. J. (1999). Magnetism from conductors and enhanced nonlinear phenomena. *IEEE Transactions Microwave Theory Techniques*, *47*, 2075–2084.

Pennec, Y., Djafari-Rouhani, B., Larabi, H., Vasseur, J. O., & Ladky-Hennion, A.-C. (2008). Low-frequency gaps in a phononic crystal constituted of cylindrical dots deposited on a thin homogeneous plate. *Physical Review B*, *78*, 104105.

Phani, A. S., & Hussein, M. I. (2013). Analysis of damped Bloch waves by the Rayleigh perturbation method. *Journal of Vibration and Acoustics-Transactions of the ASME*, *135*, 041014.

Phani, A. S., & Hussein, M. I. (2017). *Dynamics of lattice materials*. New York: Wiley.

Phani, A. S., & Woodhouse, J. (2007). Viscous damping identification in linear vibration. *Journal of Sound and Vibration*, *303*(3–5), 475–500.

Phani, A. S., & Woodhouse, J. (2009). Experimental identification of viscous damping in linear vibration. *Journal of Sound and Vibration*, *319*(3–5), 832–849.

Russell, D. A., Titlow, J. P., & Bemmen, Y.-J. (1999). Acoustic monopoles, dipoles, and quadrupoles: An experiment revisited. *American Journal of Physics*, *67*(8), 660–664.

Sigalas, M. M., & Economou, E. N. (1992). Elastic and acoustic wave band structure. *Journal of Sound and Vibration*, *158*(2), 377–382.

Smith, D. R., Padilla, W. J., Vier, D. C., Nemat-Nasser, S. C., & Schultz, S. (2000). Composite medium with simultaneously negative permeability and permittivity. *Physical Review Letters*, *84*(18), 4184–4187. https://doi.org/10.1103/PhysRevLett.84.4184.

Sprik, R., & Wegdam, G. H. (1998). Acoustic band gaps in composites of solids and viscous liquids. *Solid State Communications*, *106*(2), 77–81.

Sun, C. T., Achenbach, J. D., & Herrmann, G. (1968). Time-harmonic waves in a stratified medium propagating in the direction of the layering. *Journal of Applied Mechanics-Transactions of the ASME*, *35*(2), 408–411.

Wagner, N., & Adhikari, S. (2003). Symmetric state-space method for a class of nonviscously damped systems. *AIAA Journal*, *41*(5), 951–956.

Wang, G., Wen, X. S., Wen, J. H., Shao, L. H., & Liu, Y. Z. (2004). Two-dimensional locally resonant phononic crystals with binary structures. *Physical Review Letters*, *93*, 154302.

Woodhouse, J. (1998). Linear damping models for structural vibration. *Journal of Sound and Vibration*, *215*(3), 547–569.

Wu, T.-T., Huang, Z.-G., Tsai, T.-C., & Wu, T.-C. (2008). Evidence of complete band gap and resonances in a plate with periodic stubbed surface. *Applied Physics Letters*, *93*(11), 111902.

Yang, S., Page, J. H., Liu, Z., Cowan, M. L., Chan, C. T., & Sheng, P. (2004). Focusing of sound in a 3D phononic crystal. *Physical Review Letters*, *93*(2), 024301.

Zhang, X., & Liu, Z. (2004). Negative refraction of acoustic waves in two-dimensional phononic crystals. *Applied Physics Letters*, *85*(2), 341–343.

Zhang, X., Liu, Z., Mei, J., & Liu, Y. (2003). Acoustic band gaps for a two-dimensional periodic array of solid cylinders in viscous liquid. *Journal of Physics: Condensed Matter*, *15*(49), 8207.

Kolibri 360 GmbH
Ferdinand-Jühlke-Straße 7
99095 Erfurt, Deutschland
produktsicherheit@kolibri360.de

ज्ञान गरिमा के
68 वर्ष

भारतीय
ज्ञानपीठ

णाणं
पयासयं
वीर नि.सं. 2470
ई.सन् 1944
भारतीय ज्ञानपीठ